Martin J. Forrest

Recycling and Re-use of Waste Rubber

Also of interest

Rubber Analyis
Characterisation, Failure Diagnosis and Reverse Engineering
Forrest, 2019
ISBN 978-3-11-064027-4, e-ISBN 978-3-11-064028-1

Polymer Engineering
Tylkowski, Wieszczycka, Jastrzab (Eds.), 2017
ISBN 978-3-11-046828-1, e-ISBN 978-3-11-046974-5

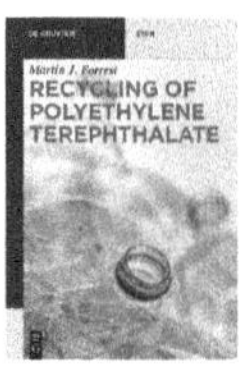

Recycling of Polyethylene terephthalate.
Forrest, 2019
ISBN 978-3-11-064029-8, e-ISBN 978-3-11-064030-4

Metals in Wastes
Wieszczycka, Tylkowski, Staszak, 2018
ISBN 978-3-11-054628-6, e-ISBN 978-3-11-054706-1

e-Polymers.
Editor-in-Chief: Seema Agarwal
ISSN 2197-4586
e-ISSN 1618-7229

Martin J. Forrest

Recycling and Re-use of Waste Rubber

2nd Edition

DE GRUYTER

Author
Dr. Martin J. Forrest
Shrewsbury Laboratory
Shrewsbury
SY4 4NR
Great Britian

ISBN 978-3-11-064400-5
e-ISBN (PDF) 978-3-11-064414-2
e-ISBN (EPUB) 978-3-11-064423-4

Library of Congress Control Number: 2018966619

Bibliographic information published by the Deutsche Nationalbibliothek
The Deutsche Nationalbibliothek lists this publication in the Deutsche Nationalbibliografie; detailed bibliographic data are available on the Internet at http://dnb.dnb.de.

Typesetting: Integra Software Services Pvt. Ltd.
Printing and binding: CPI books GmbH, Leck
Cover image: Roman_Makedonsky / :iStock / Getty Images

www.degruyter.com

Preface

This book is a review of the current state of the art and covers the main areas of interest, research and commercial exploitation that are associated with the recycling and re-use of waste rubber. It places the situation that exists today into an historical perspective and describes the various factors – environmental, legislative, economic and societal – that have helped drive innovation and research activity to the point where a very large range of potential routes for recycling rubber have now been investigated and evaluated with respect to their technical and economic merits. These include processes concerned with the production and activation of rubber crumb and powder, as well as a large number of different devulcanisation and reclaiming technologies for the regeneration of high-quality, added-value rubber products. All of the principal examples of these different technologies are covered, along with the provision of information regarding their associated advantages and disadvantages. In tandem with the increased availability of large quantities of recycled rubber there has been a considerable amount of research activity that has focused on its use to create new products, and this book also reviews the activities that have taken place in this area.

In addition to describing the technical innovations that have taken place in the main areas that are concerned with rubber recycling, this book also provides descriptions of the different quality control and evaluation methods and systems that exist for products such as rubber crumb and the large range of products in which they can be used.

The book therefore covers the full range of recycling options that are currently available, including the recovery of energy and the generation of fuel products, the recovery of important raw materials like carbon black, blending into polymeric materials, such as rubbers, thermoplastics and thermosets, and incorporation into other materials, including concrete, wood and ceramics.

In writing this book, the up-to-date, extensive information contained in the Smithers Rapra Polymer Library has been used. This has been complemented by many other sources of information, including the published findings of relevant research projects, trade literature references, and the considerable in-house expertise that has been gained at Smithers Rapra while being an integral part of the research in this field.

https://doi.org/10.1515/9783110644142-201

Acknowledgments

Many thanks to my colleagues at Smithers Information who have assisted me in the writing and production of this book, in particular Eleanor Garmson and Helene Chavaroche, who have provided advice and used their expertise with Duncan Potter from Riverside Publishing Solutions to edit the book for publication.

Thanks also go to a number of companies and organisations who have kindly provided me with permission to include their photographs and images in the book:

- Institute of Materials, Minerals and Mining, London, UK
- Watson Brown HSM Ltd, Glossop, UK
- Bob Kind of Polymer Recyclers Ltd, Dordon, Tamworth, UK
- DevulCO2/ReMould Research Consortium, http://www.devulco2.net/
- Rubber Resources B.V., Maastricht, The Netherlands
- Dr Isayev of Akron University, Akron, OH, USA
- Waste Resource Action Programme (WRAP), http://www.wrap.org.uk/
- Big Tyre Recycling Corporation B.V.B., Martelaarslaan, Belgium

https://doi.org/10.1515/9783110644142-202

Contents

1 Introduction to recycling and the re-use of rubber

Since the birth of the rubber industry, technologists and chemists have worked to develop materials and products that are highly durable, resistant to temperature and other aggressive environments and, as a consequence, give high performance and have a long life. These advantages, from the consumer's perspective, create problems when it comes to disposal at the end of a rubber product's life. The same problems arise when dealing with the normal in-process waste produced in rubber factories. Processors of thermoplastics and thermoplastic rubbers are able to recycle their moulding flash and out-of-specification products, but this has not been possible for the processors of conventional rubber.

The pressure on the rubber industry and the many users of its products to recycle rubber has increased dramatically over the last 25 years due to a combination of economic, environmental, societal and legislative factors. The need to conserve natural resources, coupled with the publication of important legislation such as the European Union (EU) Landfill Directive (1999/31/EC), which banned the placing into landfill sites of any tyre-derived waste from 1st July 2006, gave fresh impetus to the search for technologies that could render waste rubber amenable to reprocessing, impart physical properties comparable to virgin rubber, and so achieve high added value. Although some methods of devulcanising and reclaiming waste rubber have been available for a many years (indeed the earliest techniques were developed in the 19th century), most either involved breaking down the rubber mechanically into crumb, so that it could be incorporated at low levels into new rubber compounds as a diluent filler, or attempting to break the crosslinks, resulting in some degradation of the rubber, leading to a relatively low value processing aid.

The worldwide interest in rubber recycling and the important role being played by industry is apparent in the number of presentations that are being given on the subject at major conferences. For example, in a paper presented to the *Tire Technology Expo 2014 Conference*, Mergell [1] discussed the future trends in truck tyres and highlighted how new recycling and retreading technologies will help to relax tight raw material markets in a cost-effective way. This presentation also mentioned some of the technical challenges with respect to resilience, mechanical strength and scorch, which recycled rubber still needs to fully address, and the efforts that Continental are making to closing the tyre production cycle to meet their 2025 material life cycle target.

Due to these various considerations, increased funding has become available from a number of sources, including national and regional governments, and this area of research and development is now an extremely active one with a large number of recycling and re- use options being explored. It is possible to divide different

https://doi.org/10.1515/9783110644142-001

types of recycling activities and processes for waste rubber, and other polymeric products, particularly plastics, into four broad generic categories. These are:

- *Primary*: reprocessing into materials and products having properties that are the same, or at least comparable, properties to the original material or product.
- *Secondary*: where the recycled material is made into products that do not have (or need to have) properties that are the same or comparable to the original material or product.
- *Tertiary*: the reduction of the waste material or product into small chemical units (i.e., molecules) that can then be recycled into new materials and products by routes such as re-polymerisation.
- *Quaternary*: the recovery of the energy inherently present in the waste materials or products by methods such as incineration, or the burning of fuel products that are derived from pyrolysis processes.

As this book demonstrates, all of the above routes have been explored in the search for different ways to recycle and re-use waste rubber and end-of-life rubber products. Those that are at present most popular, and exploited to the highest degree, are the Secondary and the Quaternary routes. Of those processes that fall within the Secondary category, the generation of crumbed rubber, which is incorporated into a very wide range of new products (Chapter 7), is dominant at present, although devulcanisation to create a good quality product that can be added to virgin rubber at relatively high levels is increasing in popularity and the level of exploitation. With respect to the Quaternary route, its industrial importance is apparent from the very high proportion of waste tyres that are used for tyre- derived fuel (TDF), particularly in countries such as the USA.

The 'Holy Grail' of rubber recycling is to develop a 'Primary category' devulcanisation process. Such a process would be capable of recycling rubber back into new products, possessing equivalent properties in all respects to the original product, without having to use any virgin rubber in their manufacture. The attractiveness of such a process has meant that it has been one of the main goals of researchers in the field, and some good progress has been made towards achieving it, with devulcanisation processes being developed and optimised that are capable of regenerating a high quality product that has the capability to be used on its own to manufacture high-added-value products, or at very high levels of addition into virgin rubber compounds.

The challenge running alongside all this research and development (R&D) activity is to ensure that any process that achieves its technical targets will also be economically viable and so have good commercial potential. Increases in commodity prices tend to assist in achieving this commercial goal, although the costs associated with other aspects of a recycling process (e.g., energy and, if they are used, chemicals) can increase as well, reducing the overall net gain. A complete cost assessment exercise, taking into account every aspect and feature of a process (e.g., capital expenditure, material costs, labour costs, energy costs, storage, packaging and transportation costs, and so on) needs to be carried out to provide a complete and accurate evaluation of economic viability. Of course, where a product is being sold into the marketplace, the final selling price, and a robust system to establish this and vary it over time as conditions change, are also vital in ensuring that a business is profitable and viable in the long term.

In addition to the above considerations that can be applied to a single process, there are a number of ways in which the various routes that are available for recycling rubber can be compared with the costs to produce new products and materials, and one is a comparison of the energetic balance of the different processes. Some generic indicative examples have been provided by Manuel and Dierkes [2] and are reproduced in Table 1.1.

Table 1.1: Energetic balance of the production of virgin materials and recycling of rubber products.

Process	Energy expenditure (kWh/kg product)
Tyre production in total	−20
Production of synthetic polymers	−13
Production of carbon black	−13
Re-treading of tyres	−6
Incineration of waste rubber	+5
Granulating of whole tyres	−0.2
Reclaiming of rubber	−0.7

Table 1.1 shows that producing reclaimed rubber consumes only a small percentage of the energy that it takes to produce new tyres, new polymers from monomers, or even virgin carbon black. It also shows the benefits of incinerating waste rubber to generate energy and illustrates why recovering energy from waste tyres is so attractive and carried out to such a large extent around the world (Chapter 8, Section 8.4).

With regard to overall environmental impact, in order to assess if a new recycling process, or recycling option, does compare well with those that are currently being used, a full life cycle assessment (LCA) can be carried out using the framework and principals of accepted standards (e.g., International Organization for Standardization standard ISO14044). A large number of such assessments have been carried out over the last 25 years, most of them comparing the environmental impacts of recovering and recycling waste rubber as opposed to incinerating it and/or placing it in landfill sites, because they have been commissioned with a view to assisting waste management strategies. Some of the more recent ones have compared a range of options for the re-use and recycling of the waste rubber. Examples

of both of these types of recent LCA studies, all accessible in the public domain, are listed below:
- Comparative LCA of two options for scrap tyre treatment: material recycling *versus* TDF combustion [3]
- Economic advantages of using ultra-fine scrap tyre rubber in virgin and recycled rubber and plastic compounds; a comprehensive LCA [4]
- End-of-life tyre management LCA: a comprehensive analysis for Alberta Recycling Management Authority [5]

A large amount of information and data are presented in all three studies, but some of the salient points that are highlighted in their executive summaries are given below:
- The study published by Springer [3] concluded that, for the nine different environmental impact categories that were assessed, in all categories material recycling (e.g., in artificial turf products) as an end-of-life option for tyres provides a larger environmental benefit than co-incineration in cement kilns. Also, the difference between the two options in most of the categories was judged to be significant.
- The study that looked at the economic advantages of re-using waste rubber as an ultra-fine powder (100 mesh and below) in new rubber and plastic products [4] listed three areas of major economic advantage:
  - Cost savings resulting from the partial replacement of virgin natural and synthetic polymers in products
  - Savings generated by improved performance of products in use
  - Cost avoidance through lessened environment impact, specifically mitigation of petroleum use and atmospheric carbon.
- The Alberta study [5] assessed eight different options for recycling/ re-using waste tyres, ranging from crumb in sports pitches, different energy generation options, and use as a wood, asphalt, gravel, or cement replacement material. The study concluded that there was no single outright 'winner' that outperformed across the board and that for each option there were typically both net increases and decreases of varying environmental parameters. The displacement of concrete or asphalt aggregate was thought to show the most overall benefits of all the options.

Some examples of the reduction in carbon footprint that can result as a consequence of re-using rubber in a range of products and for energy generation has been published in the LCA study carried out by The Institute for Environmental Research and Education [6]. The analysis that is presented in this report was performed in conformance with applicable international standards for LCA and carbon footprinting and covered around 16% of the US tyre recycling industry. Some of the findings published in this report are highlighted below:

- The use of recycled rubber in moulded products provides a substantial carbon footprint advantage over the use of virgin plastic resins, having between four and twenty times lower carbon footprint.
- When used in road surfaces, recycled rubber has between three and seven times lower carbon footprint than asphalt on a materials basis.
- When used in energy recovery, recycled tyres provide about a 20% carbon footprint advantage over coal, but waste tyres have substantially more carbon emissions than other fossil fuels.
- The carbon footprint of tyre recycling is dominated by the processing of the tyres, followed by their transportation to the processing facility. Electricity was the largest source of this carbon footprint, with the use of diesel fuel the second largest.

The report's conclusion was that the re-use of rubber products from used tyres has the potential to make a substantial contribution to reducing carbon emissions.

In addition to delivering some of the more obvious environmental benefits that are described in the LCA studies above, other environmental benefits have been highlighted by studies that have looked into the performance of tyres that contain recycled rubber in the form of ground powder. A study conducted by Lehigh Technologies Inc. and Malcolm Pirnie Inc. [7] concluded that tyres that contain recycled rubber powder have a lower rolling resistance than standard tyres, which improves the vehicles fuel economy and so reduces the amount of greenhouse gases emitted into the environment. The results showed that incorporating 10% of rubber powder into the tread of a tyre produced a median fuel saving of 3.2 gallons/ passenger vehicle/year. The same study also showed that the presence of rubber powder reduced the air permeability of the tyre by up to 50%, which was also beneficial as tyres that are optimally inflated are safer and also roll more effectively. This maintenance of rolling resistance was estimated by translate into a median fuel saving of 0.7 gallons/passenger vehicle/year.

Often, the initial attractiveness of using waste rubber crumb (particularly tyre crumb) in products such as concrete, asphalt and wood lies in the importance of finding a use for the material itself. However, in addition to the environmental benefits mentioned above, improvements in important performance parameters can also result from its use (e.g., improvements in wear, durability, impact performance, insulation and so on), and another major benefit is that it takes the place of important raw materials (e.g., ceramics, wood, aggregate and so on) in these large tonnage products.

A number of authors have published reviews of rubber recycling and one example is that produced by Myhre and Piland [8]. Their paper covers the types of waste rubber feed stocks that are available for recycling, the types of grinding and particle size reducing equipment that are on the market, and the effect that incorporating various forms of recycled rubber (e.g., whole tyre reclaim and rubber crumb) has on the properties of a number of rubber compounds, including natural rubber (NR), ethylene-propylene-diene monomer rubber, and styrene-butadiene rubber (SBR)/ NR and SBR/butadiene rubber blends. Also, a chapter on the recycling of waste rubber, written by Srinivasan and co-workers, was included in a 2008 book entitled *Current Topics in Elastomers Research* [9]. This chapter covers all aspects of the recycling process, including the origins of the rubber, grinding and crumbing processes, the various devulcanisation processes, and the use of waste rubber in a number of applications and products, including blends with plastics, and in asphalt and concrete. In the same book, a chapter by Costa and co-workers [10] covering thermoplastic elastomers covers the creation of a type of thermoplastic rubber by blending waste rubber crumb into thermoplastic polymers, such as polypropylene. And a multi-authored book that covers all the technologies that have been developed to recycle waste rubber has been edited by De, Isayev and Khat [11] and was published in 2005 by CRC Press.

Finally, as with all recycling activities, the challenge with the recycling of rubber, in whatever form or product, has always been to develop processes that, ideally, have the following attributes:

- Efficient in operation with relatively low running costs
- Relatively low maintenance costs
- Capable of generating high quality product in a consistent manner
- Flexible design in terms of output rates
- Justifiable capital outlay
- Environmentally friendly
- Meets all health and safety requirements
- Economically competitive with existing processes and systems
- Robust with respect to satisfying changing market requirements
- Profitable!

It will be seen that meeting as many of these criteria as possible is a theme running throughout this book, and the degree to which a particular process, or end-use, addresses them is often the defining factor in their commercial viability and success.

References

1. B. Mergell in *Proceedings of the Tire Technology Expo 2014*, 11–13th February, Cologne, Germany, 2014.
2. H.J. Manuel and W. Dierkes in *Recycling of Rubber*, Rapra Review Report No.99, Rapra Technology Ltd, Shawbury, UK, 1997.
3. R. Feraldi, S. Cashman, M. Huff and L. Raahauge in *International Journal Life Cycle Assessments*, Springer Publishing, published on-line on 17th December 2012.

4. B. Bras, T. Guldberg, A.M. Cialone and P.R. George in *Proceedings of the ACS Rubber Division Meeting – Fall 2007*, Cleveland, OH, USA, American Chemical Society, Washington, DC, USA, 2007, Paper No.79.
5. G. Haines, M. McCulloch and R. Wong in *End-of-Life Tire Management LCA – A Comparative Analysis for Alberta Recycling Management Authority*, The Pembina Institute, Calgary, Canada, 2010.
6. *Carbon Footprint of USA Rubber Tire Recycling 2007*, The Institute for Environmental Research and Education, Vashon, WA, USA, on behalf of Institute of Scrap Recycling Industries Inc., Washington, DC, USA, November 2009.
7. *Tires made with Recycled Rubber Powder Improve Fuel Economy*, The Smithers Report, Ed., M.E. Fay, Smithers Scientific Services Inc., Akron, USA, 2008, **21**, 3b. [Source: Lehigh Technologies Inc., Naples, FL, USA, 4th March 2008].
8. M. Myhre and E. Piland in *Proceedings of the ACS Rubber Division Meeting – Fall 2008*, Louisville, USA, American Chemical Society, Washington, DC, USA, 2008, Paper No.18.
9. A. Srinivasan, A.M. Shanmugharaj and A.K. Bhowmick in *Current Topics in Elastomers Research*, Ed., A.K. Bhowmick, CRC Press, Boca Raton, FL, USA, 2008, p.1043.
10. F.R. Costa, N.K. Dutta, N.R. Choudhury and A.K. Bhowmick in *Current Topics in Elastomers Research*, Ed., A.K. Bhowmick, CRC Press, Boca Raton, FL, USA, 2008, p.101.
11. *Rubber Recycling*, Eds., S.K. De, A.I. Isayev and K. Khait, CRC Press, Boca Raton, FL, USA, 2005.

2 Brief history of rubber recycling

The re-use of waste rubber has been carried out to some degree by the rubber industry virtually from its inception, and the first commercial activities in rubber recycling, by reclaiming, began not long after vulcanisation was developed. Recycling rubber by producing reclaim rubber was of sufficient commercial interest in the 19th century for three different types of reclaiming process to be developed. These processes were the Thermal, Heater and Digester processes (Chapter 8, Section 8.1), which all involved heating the rubber to a high temperature with steam. The Heater and Digester processes also used chemicals (e.g., sodium hydroxide), called 'reclaiming agents'. This reclaimed rubber was a material in which both crosslinks and main-chain chemical bonds in the polymer backbone had been broken to produce a black material with a 'cheese-like' consistency that could be added to new rubber compounds as a process aid. In this role it could improve the quality of rubber extrudate by reducing shrinkage and improving the surface finish, as well as acting as a 'cheapening agent' to reduce cost. Reclaim rubber is a very effective material when used in these capacities and continues to this day to be a commercially viable product for certain sectors of the market.

Interest in rubber recycling was given a major boost by the advent of World War II, when raw materials were scarce and recycled rubber (e.g., in the form of reclaim) was required to fill the gap. Production of these types of products was therefore increased, but the reduction in the cost of synthetic and natural rubber during the 1950s resulted in a decrease in demand for reclaimed rubber. This situation was exacerbated by a number of demanding technological developments during the 1960s (e.g., steel-belted and radial tyres), which, because of the dominance of the tyre industry, had the effect of introducing additional technical and commercial limitations on the use of reclaim [1].

The situation began to show signs of change at the start of the 1970s with the oil crises, and the attendant large increase in commodity prices, and a growth in interest with respect to safeguarding the environment for future generations. In common with recycling and sustainability initiatives in other areas of science and technology, investigations into the recycling of rubber began to be of more interest and funding more readily available. In addition to investigations that looked into the recycling of waste rubber that was still in its vulcanised form (e.g., crumb rubber), researchers started to explore novel methods for the destruction of the crosslinked matrix that exists within the material (e.g., by the use of ultrasound) to increase the options available for its re-use, and to realise the potential for the manufacture of high-quality, high-value recycled rubber products that consist either solely of recycled rubber, or at least contain a very high proportion of it.

By the mid-1980s, the commercial use of recycled rubber that had been produced by the traditional methods of reclaiming was extremely small, and reclaimed

https://doi.org/10.1515/9783110644142-002

rubber represented less than 1% of worldwide rubber consumption. The effect that new environmental regulations had had on the reclaim industry was a major contributor to this decline [1].

By the mid-1990s, a large number of possible routes to achieving the goal of recycling rubber into high-value materials and products were being explored, and this activity was reflected by the number of patents that were being granted and the number of technical articles and research papers that were being placed into the public domain. Those active in the field were keen to demonstrate that the systems that they were proposing not only addressed the issues of recovering rubber that was of a high technical standard but that the recycling process itself was efficient to use and so had commercial potential and economic viability. By this time, environmental waste management regulations had resulted in some major industrialised countries (e.g., Germany) having targets as high as 100% for energy and material recycling rates for products such as waste tyres [2]. The recycling of tyres was also being pursued very actively in other areas of the world; for example, Japan had an energy recovery rate of 43% from waste tyres, with only 8% being disposed of [3], and the USA was utilising 76% of waste tyres, with 57% being incinerated for energy recovery [4].

In recent years, a major impetus for increased interest and research activity into recycling and re-using waste rubber has been the advent of environmental legislation such as the European Union (EU) Landfill Directive (1999/31/EC) (European Community), which initially banned the placing of whole tyres into landfill sites from 1st July 2003 and finally banned the placing of any tyre-derived waste into landfill after 1st July 2006. Another important piece of legislation was the End-of-Life Vehicle Directive (2000/53/EC), which set escalating targets for the recycling of components (of all material types – rubber, plastic, metal and so on) from vehicles (Chapter 3). This legislation, and others like it in the EU and other parts of the world, came about as governments realised that the pressures on the environment were such (e.g., the finite capacity of landfill sites in the UK) that mandatory action to force change was required. It was also hoped that the legislation would stimulate activities that would generate economic opportunities and benefits within the areas in which it was implemented.

As a consequence of this legislation, the continued increase in environmental awareness throughout society, and the maintenance of high commodity prices, an exponential increase in initiatives targeting the recycling and re-use of waste rubber has taken place. A good illustration of this is provided by the many research projects and recycling initiatives that have been funded, and carried out in partnership with industry and local government, by the Waste Resources Action Programme (WRAP) in the UK. Due to their diversity and pertinence, a number of these WRAP-funded activities are highlighted in this book.

Another illustration of the impact of these influences, such as increasing landfill taxes for general rubber goods products, is that more forms of waste rubber are being considered for recycling. One example of this is the recycling of post-industrial and post- consumer ethylene-propylene-diene monomer (EPDM) rubber membranes, which has received a great deal of attention in recent years as described in an article published in *Rubber and Plastics News* [5].

An optimistic article on the future of rubber recycling, particularly of higher-value rubbers such as butyl rubber, was published by the Gujarat Reclaim and Rubber Products company of India in 2009 [6]. In this article the company states that recycling butyl rubber from sources such as waste inner tubes is attractive due to the relatively high cost of the rubber and the fact that it is relatively easy to separate this type of rubber waste from other rubber products, hence avoiding a major source of contamination. The company also claims that in addition to being environmentally friendly, such an activity is economically viable, as one tonne of recycled butyl rubber has a value half that of the virgin material. It can therefore provide a reasonable return for the company selling it as well as potentially increasing profit margins for a company using it.

With regard to the recent waste tyre situation in Europe, by 2010 the European Tyre & Rubber Manufacturers' Association (ETRMA) was able to claim that the market for tyre-derived products was 'sustainable and economically viable' and that the European recovery rate for end-of-life tyres had reached 96% [7]. In the same article the UK was reported as treating 100% of its end-of-life tyres. The figures that were included showed that of the 479,000 tonnes of end-of-life tyres arising in the UK during 2009, 8.5% were retreaded, 45.1% were processed into other materials, 24% were used as a source of energy, and 19% were re-used or exported.

One illustration of how increasing the quantity of recycled rubber being re-used is becoming an important part of corporate strategies in 2014 for increased sustainability is the recent report in *Scrap Tire News* describing how Ford Motor Company researchers have led a study to determine a formulation strategy to incorporate products from recycled ground tyres and recycled EPDM into virgin formulations for use in automotive parts [8]. The study examined the incorporation of recycled materials (ground rubber powder, devulcanised rubber and recovered carbon black) into EPDM recipes as both carbon black replacements and inert fillers. The results obtained were sufficiently encouraging for a recommendation that the use of recovered carbon black is pursued as a replacement for virgin carbon black and that the devulcanised rubber and rubber powder continue to be considered for use as inert fillers. The article also reports that future research work will include the use of recycled materials in tyre treads.

References

1. H.J. Manuel and W. Dierkes in *Recycling of Rubber*, Rapra Review Report No.99, Rapra Technology Ltd, Shawbury, UK, 1997, p.3.

2. *ETRA Newsletter*, European Tyre & Rubber Manufacturers' Association (ETRMA), Brussels, Belgium, Autumn 1995, p.13.
3. K. Nishimura, *Kautschuk Gummi Kunststoffe*, 1993, **46**, 12, 989.
4. *Scrap Tyre Use/Disposal Study – 1996 Update*, Scrap Tyre Management Council, Washington, DC, USA, April 1997.
5. B. Dawson, *Rubber and Plastics News*, 2009, **38**, 4, 12.
6. D. Shaw, *European Rubber Journal*, 2009, **191**, 3, 39.
7. Anon., *Tyres and Accessories*, 2010, **9**, 42.
8. J.L. Tardiff, P.R. Rohweder, C.M. Flanigan and M. Harper, *Scrap Tire News*, 2014, **28**, 3, 1.

3 Overview of the world rubber recycling market

To help appreciate why so much research and development (R&D) activity has been targeted at recycling tyres it is informative to look at a few fundamental statistics. In a recent presentation by Sekhar [1], the following figures were shown that highlight the global problem that waste tyres represent today:

- 7 billion people in the world
- 1.1 billion vehicles on the road
- 1.7 billion new tyres produced a year
- Over 1 billion waste tyres generated per year

Although the rubber industry is a very mature and complex sector, with over 25 different generic types of rubber (e.g., natural rubber (NR), styrene-butadiene rubber (SBR), nitrile, ethylene-propylene- diene monomer (EPDM) rubber, fluorocarbon rubber, silicone and so on) being used in the production of numerous types of end- product (hoses, seals, gaskets, diaphragms and so on) for a myriad of applications and end-uses, no single rubber product comes near to dominating the market as tyres do. It is for this reason, and for convenience, that the rubber industry is frequently split by analysts and commentators into two distinct sectors: the tyre sector and the general rubber goods (GRG) sector. The GRG sector encompasses all other rubber products. The dominance of tyres that is expressed above is apparent in the division in the total world consumption of rubber between the tyre and GRG sectors, which is approximately 50:50.

Other global statistics have been used to provide simplified life cycle assessment descriptions for the rubber industry. For example, Brown of Watson Brown HSM Ltd [2] has provided the following global estimates for recycling:

- Total amount of rubber recycled at its end-of-life: typically 3–15%
- Amount of waste rubber re-used in some way (e.g., retreading, new products and so on): 5–23%
- Amount of waste rubber consumed for energy recovery: 25–60%
- Amount of waste rubber sent to landfill or stockpiled: 20–30%

Another global figure shown by Brown in this presentation was that factory waste accounts for 5–15% of total production, showing the need for and attractiveness of in-house closed-loop systems, a market that the high stress mixing (HSM) technology that Watson Brown HSM Ltd developed for re-using waste rubber has addressed (Chapter 4, Section 4.4.2). A specific mention was also made by Brown of the amount of rubber that is produced as scrap by some typical rubber manufacturing processes. For example:

https://doi.org/10.1515/9783110644142-003

- Stamping out of gaskets from cured sheet rubber: 20–40% waste
- Injection moulding processes due to moulding ‘flash’: 30–50% waste
- Extrusion processes to produce final product: 2–5% waste

With regard to consumption, although accurate figures for rubber materials and products are hard to obtain for the world economy due to the complexity of the market and the fact that economic factors play such an important role in determining the market price that exists for the raw ingredients at any one point in time, rough estimates place the split in the consumption of NR and all synthetic types of rubber at approximately 40:60, with the dominant synthetic rubber being SBR. This 40:60 split is reflected in figures that were released for the Chinese rubber industry in 2010, which showed the consumption of NR at 2.89 million tonnes and that of synthetic rubber at 3.41 million tonnes [3].

The large amounts of NR and SBR consumed by the world economy is due to the fact that they are the main ‘tyre rubbers’, with the tread compounds of tyres (particularly those of passenger vehicles and trucks) usually being sulfur-cured compounds based on these two diene-type rubbers, sometimes with an amount of another diene rubber (polybutadiene) added to improve properties such as flex crack resistance.

In general, the smaller a tyre the larger the proportion of SBR will be, whereas as tyres get larger the amount of SBR diminishes, mainly to reduce hysteresis and the consequent heat build-up during service. The tread compound of an ordinary passenger car tyre will therefore be manufactured using a blend of SBR/NR at a ratio of around 60:40, a truck tyre will have an SBR/NR ratio of 30:70, and a large tyre for an earth-moving construction vehicle from NR only.

With respect to the amounts of waste that are generated by end-of- life tyres, it has been estimated that in the USA around 3.3 million tonnes of tyre waste are generated per year, whereas the figure in the European Union (EU) economic area is 2.5 million tonnes, and in the UK alone it is around 0.5 million tonnes [4].

Looking more closely at the fate of tyre waste in the USA, it was estimated in 2011 [5] that only around 10% was being recycled into new products, with over 50% being burned for energy recovery (e.g., tyre-derived fuel (TDF) oil) and the rest being discarded into landfill and so on. With respect to the proportion of tyres that are placed into landfill worldwide, a recent estimate has put this at 25% of the total number of waste tyres [6].

At the Tyre Recovery Association annual forum in 2011, the combination of successful partnerships, participation and proactive involvement from many organisations across the tyre and reprocessing industries was cited as being critical to the success of the UK’s tyre recycling programme, which had enabled the country to continually meet the demands of the EU’s Landfill Directive. The delegates at this meeting were informed that the UK managed to reprocess around 480,000 tonnes of waste tyre materials in 2009. This compared very favourably with the rest of the EU, where only Germany, with a larger automotive sector, recycled more tyres in the same year, at around 570,000 tonnes [7]. The European Tyre & Rubber Manufacturers' Association (ETRMA) have published figures that provide a breakdown of the fate of these UK waste tyres. These figures show that, of the 480,000 tonnes, 8.5% tyres were retreaded, 45.1% were reprocessed into other materials and products, 24% were used as a source of energy, and the remainder were exported or reused in some way [8].

Although, to date, some of the most significant initiatives to reduce the environmental impact of waste tyres have been taken in Europe, other parts of the world are also addressing this issue. For example, in September 2010, China's Ministry of Industry and Information published a new strategic policy document that set out the future of the country's tyre industry. It covered all parts, from encouraging investment in indigenous NR production to improvements in energy efficiency in tyre factories, and to the responsible disposal of waste tyres [9]. In the Russian Federation the situation has been addressed in reports that have highlighted the fact that there are almost no legal regulations for handling waste materials, such as automobile tyres, as a source of secondary resources. Such a report was published in 2012 [10] and analysed the legal regulations for handling waste and highlighted the need for action to be taken in the recycling of waste tyres. It also looked at the possibilities for recycling the non-rubber components within tyres (e.g., the fibres and steel cord) in the civil engineering and construction industry, and suggested technological and economic measures for recycling tyre components into products such as tiles and fibre-reinforced concrete. In the provinces of Canada, customers who buy new tyres pay a fee of around Can$5 per tyre, which goes towards the cost of recycling waste tyres into a range of products varying from running tracks to roof shingles. Ontario, which generates around 12 million waste tyres each year, was the last jurisdiction in Canada to adopt such a scheme in 2008 [11]. In the USA the retreading industry plays a significant part in the rubber recycling sector. In 2011 there were 830 high-tech retreading plants in the USA, producing more than 15 million retreads annually [12]. Retreads account for a little more than a half of the replacement tyres used in the truck industry, ambulances and other emergency vehicles, but around 80% of the tyres used in the aircraft industry. The fact that a retreaded tyre can cost only 25% of the price of a new tyre is helping to sustain this market, as well as having the added benefits of reducing the cost of solid waste disposal and conserving valuable natural resources.

A piece of EU legislation that has had an important impact in encouraging the recycling of rubber has specifically addressed the automotive sector, which, in conjunction with the transportation sector in general, is estimated to account for around 70–75% of all rubber articles produced [13]. This legislation is the End-of-Life Vehicle Directive (2000/53/EC) and, in recognition of the large contribution that end-of-life vehicles make to the use of landfill space, it set escalating targets for the recycling of car components, culminating in a final target of 95% for all car components, to be achieved by 1st January 2015. Serious efforts in this direction are also being made in other ways, such as the Vehicle Recycling Partnership cooperation between Ford, General Motors and Chrysler, which has worked towards establishing the most efficient way to dismantle and recycle automotive components.

Papers published by workers at the Toyota R&D Laboratories [14, 15] have provided the estimates shown in Table 3.1 of the proportions of different generic types of rubber products, by weight, in a modern car.

Table 3.1: Proportions of generic rubber products in a modern car.

Rubber product	Approximate weight (%)
Tyres	65
Weather strip	10
Vibration isolators	7.5
Hoses	7.5
Miscellaneous	10

Table 3.1 highlights the importance of finding ways of recycling tyres from end-of-life cars and also shows why weather strip, which has the benefits of being a relatively easily accessible and collectable product manufactured from a single rubber type (EPDM), has also been targeted by a number of research workers and industry initiatives. The same paper also provided the approximate splits shown in Table 3.2 for the different rubber types.

Table 3.2: Proportions of generic rubber types in a modern car.

Rubber type	Approximate weight (%)
EPDM	50
NR	17
SBR	13
Polychloroprene	10
Miscellaneous	10

A number of new automotive rubber components are now on the market that contain a portion of recycled rubber. These include tyres, where around 5% can be found in the tread compound, and products in static applications that are used to insulate and seal.

In Europe, the major tyre manufacturers have formed 'Producer Responsibility' bodies to collect and recycle their scrap tyres and they fund this process with an up-front charge on their customers' invoices [16]. This charge then funds a national free recovery network, but there are difficulties, because not all tyres and tyre brands are included in these programmes.

One way of improving recycling rates in the automotive sector, as well as other sectors, is through design. A recently patented process [17, 18] has replaced the metal carrier used in a number of automotive sealing profiles with a thermoplastic one that is inserted into the profile in the extruder head by means of a co-extrusion process. The use of a highly filled polypropylene compound in combination with a low-hardness EPDM compound is claimed to be the most cost-effective combination for the metal replacement strip, although it is also possible to use a thermoplastic rubber in place of the EPDM. In addition to the improved recyclability, the new product is also claimed to be lighter and cheaper to produce.

As mentioned above, tyres are manufactured by vulcanising diene rubbers using sulfur-based cure systems. It is for this reason, in conjunction with regional factors such as the landfill ban on tyre waste within the EU in July 2006 (Chapter 2), that efforts to devulcanise rubber have concentrated on sulfur-cured diene rubbers, particularly the main tyre rubbers (NR and SBR), and tyre waste in the form of crumb or powder is often the starting material for these types of studies. With regard to devulcanisation, it is the case that the properties that the sulfur–sulfur bonds impart to the diene rubber matrix mean fortuitously that it is potentially easier to successful target the crosslinks in these types of rubbers (Chapter 4, Section 4.2). In the case of a number of other crosslinking systems (e.g., peroxide) there is much less difference between the chemical, physical and biological properties of the crosslinks and the main chain bonds, and this can cause problems when it comes to devulcanising these materials.

One of the major problems facing any organisation that is trying to develop a novel devulcanisation process, or run an existing one on a profitable basis, is the wide fluctuations that can occur to the market price of rubber, particularly NR and SBR rubber. Fundamental economic factors such as the crude oil price, and supply and demand, can affect the price of rubber. For example, because of the global economic recession and other factors, such as a rationalisation of tyre capacity, North American consumption of SBR dropped by about 19% between 2005 and 2009 [19], and this period also saw a decrease in the price of SBR and NR, so achieving a satisfactory degree of economic viability for recycling processes was made harder. However, from 2010 onwards the demand for tyres, and hence SBR and NR, increased, and so did the price.

An additional influence on the price of NR is the number and output of plantations. Recent years have seen this factor playing an important role, as some main suppliers of NR have switched some land use from this crop to others such as palm oil, and this has coincided with an increase in demand for NR from countries such as China as their automotive industry expands [20]. This factor in determining the NR price is hard to manage in the short term as it takes around

10 years for a new plantation to reach maturity. This has also been commented on by Damardjati and Jacob [21], who mention that, despite an increase in re-planting and the creation of new plantations in response to this higher demand, supply of NR will remain constrained during the next decade due to the lower yields expected from the non-traditional areas within the major NR- producing countries and the significant loss of older trees that has taken place. They also report on the average annual growth of NR in the major producing countries as being 0.9% during the period 2006–2010, with the average yield reducing by -3% and annual production increasing by 0.7%.

The expression 'added value' is often encountered in initiatives and activities that seek to re-use waste rubber in new products. An illustration of how this can be put into practice was provided in a presentation to the *RubberCon 2014* (Manchester, UK) by Kind [22]. The presentation concerned the use of tyre crumb as a major ingredient in retread compound, and some up-to-date and informative price comparisons were provided that showed the potential cost benefits of using waste rubber crumb to manufacture new products. The figures provided are shown in Table 3.3.

Table 3.3: Price comparisons of options for end-of-life tyres [22].

Product/service	Cost per tonne (£)
Gate fee for end-of-life tyres in UK	50
Tyres intended for tyre-derived fuel	50
Truck tyre crumb (40 mesh)	250
Reclaimed rubber	600
Retread rubber compound	2,500

Table 3.3 shows that the value of tyres for TDF is only around 2% of the original cost of the rubber compound and that of crumb only around 10%. Because it is relatively cheap, if it is possible to incorporate relatively large amounts of crumb into tyre retreading compound, significant savings can be made. It was demonstrated how the addition of rubber crumb to a tyre tread compound could affect its price. The information provided is shown in Table 3.4.

In order to realise these savings it is necessary, of course, to be able to produce a high-quality tread compound that is acceptable to the market by this route. Accordingly, Polymer Recyclers Ltd have carried out research, supported by Newgen Recycling Ltd and the UK Technology Strategy Board, to see if truck tyre crumb can be used at high levels (e.g., 40%) in retreading compounds for OTR, agricultural and solid tyre applications. It was reported to *RubberCon 2014* that the findings of the work that had been carried out to date indicated that it was highly likely that,

Table 3.4: Price savings from using rubber crumb in retreading compound [22].

Tyre retreading compound (phr)	Rubber crumb (phr)	Price of retread compound (£/tonne)	Saving (%)
100	0	2,000	0
90	10	1,825	8
80	20	1,650	17
70	30	1,475	26
60	40	1,300	35

by the end of the project, this would be shown to be the case. An OTR tyre that had been retreaded with a compound that contained 40% of truck tyre 40 mesh crumb as part of the work carried out during the project is shown in Figure 3.1.

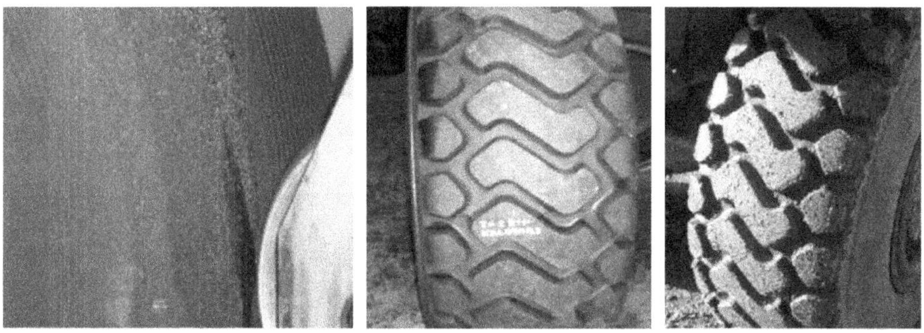

(a) (b) (c)

Figure 3.1: Off-the-road (OTR) tyre retreaded with compound containing 40% truck tyre 40 mesh crumb: (a) tyre carcass being retreaded with the compound; (b) moulded tyre ready for use; and (c) tyre fitted onto the OTR vehicle. Reproduced with permission from Polymer Recyclers. ©2014, Polymer Recyclers Ltd.

References

1. G.B. Sekhar in *Proceedings of the Tire Technology Expo*, Cologne, Germany, 2014.
2. D. Brown in *Proceedings of the Recycling of Rubber Meeting*, Institute of Materials, London, UK, December 2008.
3. P. Venugopal, *Rubber Asia*, 2010, **24**, 3, 73.
4. P.W. Dufton in *End-of-Life Tyres – Exploiting their Value*, Rapra Industry Analysis Report Series, Rapra Technology, Shawbury, UK, 2001.
5. *Polymers & Tyre Asia*, 2011, April/May,80.
6. *The Rubber International Magazine*, 2010, **12**, 10, 14.
7. *British Plastics and Rubber*, 2011, September, 47.

8. *Tyres and Accessories*, 2010, **9**, 42.
9. D. Shaw, *European Rubber Journal*, 2010, **192**, 6, 13.
10. V.S. Demyanova and A.D. Gusev, *Construction and Architecture*, 2012, **4**, 4, 28.
11. *Ontario Tire Fee to Fund Recycling*, Ed., M.E. Fay, Smithers Report, 18th October 2008, **20**, 35D, 18, Smithers Scientific Services Inc., Akron, USA (Source: *The Toronto Star*, Ontario, Canada, 28th March 2008).
12. P. Fisher, *Rubber and Plastics News*, 2011, **41**, 3, 28.
13. W.H. Klingensmith and K.C. Baranwal in *Handbook of Elastomers*, 2nd Edition, Eds., A. Bhowmick and H.L. Stephens, Marcel Dekker, New York, NY, USA, 2001, p.834.
14. K. Fukumori, M. Matsushita, M. Mouri, H. Okamoto, N. Sato, K. Takeuchi and Y. Suzuki, *Kautschuk Gummi Kunststoffe*, 2006, July/August, 405.
15. K. Fukumori and M. Matsushita, *R&D Review of Toyota CRDL*, 2003, **38**, 1, 39.
16. P. Taylor, *Rubber Asia*, 2012, **26**, 3, 26.
17. D. Shaw, *European Rubber Journal*, 2010, **192**, 5, 35.
18. F. Reinecke, *Rubber Fibres Plastics International*, 2010, **5**, 4, 166.
19. I. Lerner, *ICIS Chemical Business*, 2011, **279**, 3, 30.
20. P. Venugopal, *Rubber Asia*, 2010, **24**, 3, 73.
21. D.S. Damardjati and J. Jacob, *Rubber Asia*, 2010, **24**, 6, 74.
22. R.C. Kind in *Proceedings of RubberCon 2014*, 14–15th May 2014, Manchester, UK, IOM Communications, Institute of Materials, Minerals and Mining, London, UK, 2014.

4 Review of devulcanisation techniques

4.1 Introduction

As mentioned in Chapters 2 and 3, a number of factors are driving the need for rubber recycling and it is a very active area of research. As a consequence, a large number of different technologies have been developed for the devulcanisation of waste rubber and this high level of activity is reflected in the number of articles, books and reviews that have been published over the last 25 years. Of the reviews that have been published, two major examples that cover both the devulcanisation technologies that are available, and rubber recycling in general, have been published in the journal *Rubber Chemistry and Technology* [1, 2].

In addition to reviews being generated by research scientists, consultancy companies have been commissioned to carry out reviews. A prominent example of this type of review is the major appraisal of tyre devulcanisation technologies that was undertaken by CalRecovery Inc. for the Integrated Waste Management Board of California [3]. The resulting report, which was published in December 2004, provided a comprehensive description of each technology and the research and development (R&D) that was being undertaken, as well as having sections that addressed cost, market and environmental analysis. The report identified approximately 25 potential devulcanisation technology researchers and developers throughout the world and North America. However, it concluded that, at the time, only a very small number of devulcanisation processes were operational and that these were primarily small-capacity systems that devulcanised both natural rubber (NR) and synthetic rubbers. The situation has changed significantly over the last 10 years due to the factors outlined in Chapters 2 and 3. These have led to a greater degree of public and commercial interest in the recycling of waste rubber and this has been reflected in the amount and diversity of the research that has taken place in the area of devulcanisation.

This section will review the different technologies that are currently available for the devulcanisation of waste rubber from waste tyres, or from products in the general rubber goods (GRG) sector. Devulcanisation, although important, is only one way of recycling waste rubber and the extensive range of other technologies that have been developed for this purpose are covered in later sections of this book; e.g., crumb manufacture and the use of crumb in the production of a variety of different products are reviewed in Chapters 6 and 7.

Trying to divide up the different technologies that exist for devulcanising rubber into discreet categories can be problematic as, no matter what system is devised, there are often 'grey' areas between some of the generic classes. As will be seen later, it is also the case that not all the approaches that have been explored were ever intended to be commercialised, and some are deliberately used on a

https://doi.org/10.1515/9783110644142-004

small scale where the process can be tightly controlled and monitored, which enables the study of the influence of different chemical devulcanising agents or reaction conditions. The category that contains the most examples of these types of processes is the Thermal category (Section 4.3).

The principal generic types of devulcanisation technology that are most commonly encountered in industry and in the literature can be summarised as shown below in Table 4.1. The generic categories shown in Table 4.1 have been used as the basis for deciding upon the structure of this section of the book, with the exception that the 'thermal' and 'thermal with chemicals' categories have been combined into the thermal section (Section 4.3). Examples of devulcanisation technologies that are regarded as not fitting perfectly into any single category have been placed together in the miscellaneous section (Section 4.9).

A brief description of each of the principal devulcanisation processes shown in Table 4.1 is provided in Sections 4.3 to 4.8, along with a summary of their strengths and weaknesses. Also, for each type of process, a review of the literature has been carried out to provide a selection of the organisations, researchers and companies that have carried out investigations and studies in that particular field.

Table 4.1: Generic categories of devulcanisation technology.

Technology	Basis of process
Thermal	Heat-induced crosslink scission
Thermal with chemicals	Targeted chemical reactions at elevated temperature
Mechanical[a]	Shear-induced crosslink scission
Mechanical with chemicals[a]	Shear/chemical devulcanisation
Ultrasonic	Ultrasound energy
Microwave	Energy generated by microwaves
Microbiological	Microorganisms

[a]These processes can either be operated at, or near, ambient temperature, or heat can be introduced into the process

It is not possible to provide a completely exhaustive or accurate list of the companies that are actively involved in devulcanisation because, as with any complex commercial sector, the situation is changing constantly, with companies entering and leaving the market, or changing the technology that they use. However, a selection of some of the organisations and companies that have been encountered during the writing of this book, and are believed to be, or have been, involved in this area of work from the literature that is in the public domain, is provided in Table 4.2.

In addition to the two reviews that are mentioned at the beginning of this section, a number of other relatively recent reviews of the processes and methods that

Table 4.2: Organisations associated with the devulcanisation of waste rubber.

Organisation	Location	Type of process[a]
BF Goodrich Company	USA	Chemical/mechanical
Exxon	USA	Chemical/mechanical
Goodyear Tire & Rubber Company	USA	Range of processes
Green Rubber	USA	Chemical/mechanical
Revultec	USA	Chemical
University of Akron	USA	Ultrasonic
Redwood Rubber	USA	Ultrasonic
NRI Industries	Canada	Mechanical
Polymer Recyclers	UK	Mechanical
Watson Brown HSM	UK	Mechanical
H S Polymer Reprocessing	UK	Mechanical
RecyclaTech	UK	Microbiological
Revulcon	Germany	Mechanical
Marangoni	Italy	Thermal
Rubber Resources	Netherlands	Chemical
Elast-Tekhologii	Russia	Thermal/mechanical
UAB Devulco/Institute of Chemistry	Lithuania	Shear/chemical
Levgum	Israel	Chemical/mechanical
Sumitomo Rubber Industries	Japan	Microbiological
Toyota R&D Labs	Japan	Thermal/mechanical
SRI Elastomers	Malaysia	Proprietary activation system

[a]This column only provides an indication of the type of process. For example, those termed 'mechanical' will also often involve heat as part of the process being used

have been developed and evaluated for the devulcanisation of waste rubber are available. For example, Majumdar published an overview in 2009 in the *Chemical Weekly* journal [4], which covered the challenges that face workers in this field and covered the main types of systems that have been developed (i.e., chemical, microwave, ultrasonic and so on). This article also covered the production and use of rubber crumb from waste rubber. In another article [5], Majumdar reviews the three main sources of reclaimed rubber that are available in the marketplace (rubber crumb, rubber powder and chemically digested reclaimed sheet) and describes their properties and uses.

4.2 Structure of sulfur vulcanised rubber and the properties of sulfur crosslinks

Because of the dominance of sulfur-cured rubbers in the market, which is principally due to their use in manufacturing the single most important rubber product (i.e., tyres), the majority of the work that has been done on devulcanisation has concentrated on breaking the sulfur–sulfur crosslinks in these type of rubbers, and this is reflected in the information that is presented in Sections 4.3 to 4.9. Before the different devulcanisation technologies are described in detail, it is useful to consider the chemistry that is associated with these type of crosslinks, as it is this that has played such an important role in determining the fundamental choices that have been made with regard to the mechanisms that have been employed and the development work that has been carried out.

When a rubber is subject to vulcanisation, with sulfur or any other curing agent, it is converted into a thermoset system that is lightly crosslinked, with a molecular weight (Mw) between each crosslink of approximately 10,000 Da. This leaves the majority of the rubber chain free and, as a consequence, the glass transition temperature of the material is below ambient temperature. This gives rubber its unique properties, i.e., excellent flexibility, high elasticity and very good energy absorption. When a rubber is cured using a typical sulfur-based cure system, the rubber matrix contains the following principal types of crosslinks (x-links):

- Monosulfidic x-links: C–S–C
- Disulfidic x-links: C–S–S–C
- Polysulfidic x-links: C–Sx–C

(Note: Sx where $x > 3$).

Ideally, any devulcanisation system has to attempt to remove all three types of crosslinks from the rubber matrix as rubber cannot be re-processed into a high-quality product until a high proportion (e.g., >80%) of the crosslinks are broken so that the polymer chains can flow freely past one another.

In order to develop processes and techniques capable of devulcanising sulfur-cured rubbers, polymer chemists have often referred to a number of fundamental studies that have been carried out in the past to characterise the nature and properties of sulfur bonds. These original studies, carried out by workers such as Tobolsky [6], and Murakami and Ono [7] among others, highlighted mechanisms and routes (e.g., exploitation of the relative weakness of the sulfur–sulfur crosslink bonds) that had the potential to selectively cleave the sulfur crosslinks without damaging the rubber molecules themselves. The research work that has resulted showcases the existence of three principal mechanisms that can be utilised in the devulcanisation of sulfur-cured rubbers. These three mechanisms are associated with differences that exist in the three fundamental properties of the different chemical bonds that are present in the sulfur crosslinks and the rubber molecules, namely:

- Differences in the bond energies
- Differences in the shear stiffness
- Differences in the chemical reactivity

In addition to these three properties, the other property of the crosslinks in sulfur-cured rubbers that has been successfully exploited to achieve devulcanisation is the greater susceptibility of the sulfur atoms within them to biological attack by certain bacteria (i.e., chemolithiotrophic bacteria).

A brief summary of each of the four properties referred to above is provided below.

4.2.1 Covalent bond energies

There is a difference in the bond energies that are associated with the different types of bonds within the crosslinks and the main- chain carbon–carbon bonds that exist within the rubber molecules themselves. These bond energies are shown in Table 4.3 [8].

Table 4.3: Bond energies in sulfur crosslinks and rubber molecules [8].

Bond type	Bond energies (kJ/mol)	Location in crosslinked rubber
S–S	270	Di- and polysulfidic crosslinks
C–S	310	Monosulfidic crosslinks
C–C	370	Main-chain bonds in rubber molecules

From the above, it is apparent that carbon–sulfur and sulfur–sulfur bonds have lower dissociation energies, and so will break more easily than the main-chain carbon–carbon bonds in the rubber molecules when heat is applied to the crosslinked rubber.

4.2.2 Stiffness of the covalent bonds

There is also a difference in the 'stiffness', or 'elasticity', of these different types of bonds and the relative values of their elastic constants are shown in Table 4.4 [8].

Table 4.4: Elastic constants of bonds in sulfur crosslinks and rubber molecules [8].

Bond type	Elastic constant	Location in crosslinked rubber
S–S bonds	~3	Di- and polysulfidic crosslinks
C–S bonds	Intermediate value	Monosulfidic crosslinks
C–C bonds	~100	Main-chain bonds in rubber molecules

Table 4.4 shows that the sulfur–sulfur and sulfur–carbon bonds are ‘stiffer’ and less ‘elastic’ than the main-chain carbon–carbon bonds in the rubber molecules and so will tend to break more easily upon the application of a shear force to the crosslinked rubber.

4.2.3 Reactivity of the covalent bonds

Due to their underlying chemical properties and atomic structures, there is also a difference in the chemical reactivity of the different types of bonds that are present in the rubber matrix and so if chemical agents (sometimes referred to devulcanising agents in the literature) are used that specifically target the sulfur–sulfur and sulfur–carbon bonds, usually aided by other agencies such as heat, this difference in reactivity can be exploited to achieve devulcanisation.

Because of the nature of rubber processing equipment (i.e., mills and extruders), it is usually the case that two or more of the mechanisms described above are occurring at the same time. For example, heating or shearing the rubber will rupture the sulfur–sulfur crosslinks and then chemical agents can be employed (e.g., disulfide compounds) that react with the free radicals that are formed to stop them recombining.

4.2.4 Microbiological attack on the sulfur atoms

The sulfur atoms in the sulfur crosslinks are susceptible to microbiological attack by chemolithiotrophic bacteria. These bacteria selectively consume the sulfur atoms, which leads to desulfurisation of the rubber and a corresponding reduction in the crosslink density. The bacteria have to have intimate contact with the sulfur atoms, and they are not able to migrate through the rubber to any significant degree, so the effect is usually localised at the rubber surface. Presenting as large an area as possible to the bacteria (e.g., by the use of very fine rubber powder) is therefore advantageous to the effectiveness of the process.

It can be seen from Sections 4.2.1 to 4.2.4 that four generic categories of ‘difference’ are available to research scientists that can be used as the basis for a devulcanisation process. Some processes mainly use just one mechanism, but it is often the case that in order to develop an optimised system that has the potential to be used commercially, more than one process is used, and often all three of the mechanisms that are described above are employed. Processes of this type are included in Section 4.5 and employ shear and a degree of heat, using either a two-roll mill or an extruder, and chemical agents.

As will become apparent in the sections below, these inherent differences between the bonds in the rubber molecules and those in the sulfur crosslinks are available to be used, but unfortunately they are not so great as to make the task of selectivity easy, and the challenge has always been to develop a process that finds the correct balance between efficient destruction of the crosslinks whilst inflicting minimal damage to the rubber molecules.

4.3 Thermal devulcanisation processes

4.3.1 Background to process

The types of processes that are reviewed in this section are those that have an elevated temperature as a common denominator and, by the use of this, they target the thermally labile nature of the sulfur – sulfur bonds in the crosslinks. There are quite a few examples where chemical agents have also been used in the process to assist in the devulcanisation of the rubber. In these cases, the chemical agents that have been chosen have often been similar to those that are used in the mechanical–chemical processes (Section 4.5) and often their assimilation into the rubber matrix has been assisted by solvents that have a high affinity for both the rubber and the chemical agents. Other processes of this type have employed solvents (e.g., supercritical carbon dioxide (CO_2), supercritical water, alcohols and so on) on their own, without any other chemicals, and sometimes they have reacted with the crosslinks and/or the polymer chains.

A popular reaction vessel to carry out this type of work is an autoclave. Because of its low cost and good health and safety credentials, supercritical (or liquid) CO_2 is often used, although its nonpolar nature means that it has its limitations as a solvent, particularly for certain chemical agents. In common with the majority of the other processes that are reviewed in this section, the waste rubber is usually in the form of crumb or powder to present a large surface area to the solvent/reactants in the autoclave.

Once the reaction vessel is charged with waste rubber, solvent and chemical agents, it is sealed and then heated, often at a relatively high temperature (e.g., 180 °C for 1 h), to achieve the devulcanisation. High levels of devulcanisation (e.g., up to 100%) have been claimed for gum stock rubbers, but lower levels are usually obtained with compounded rubbers, e.g., tyre rubber.

The main drawbacks to this type of process as far as commercial exploitation is concerned are that they are batch processes, usually on a small scale, and the devulcanisation reaction can take a relatively long time to reach completion. They can therefore be very inefficient compared to large-scale continuous processes. However, although their inherent limitations place restrictions on their commercial viability, this type of process has the advantages of being able to be used relatively quickly, on a small scale, in a well-structured and controlled way, and so it is a valuable tool for investigating the influence that changing certain parameters (e.g., time, temperature, chemical agent concentration, chemical agent type and so on) can have on a devulcanisation process before an attempt is made to scale it up.

4.3.2 Examples of processes and published studies

Chen and co-workers [9] used both supercritical CO_2 and supercritical water to devulcanise and depolymerise 30 mesh tyre rubber and pure, uncompounded Malaysian ribbed smoked sheet (i.e., RSS grade) NR in an autoclave. The products produced were characterised using a number of chemical analysis techniques. Data obtained by gel permeation chromatography (GPC) showed that relatively low Mw products (Mw ≈ 1,000–10,000 Da) were obtained, indicating that both crosslink scission and main-chain scission were occurring, with the chain length reducing with reaction time. The results also showed that a higher degree of main-chain breakage (i.e., depolymerisation) occurred when supercritical water was used, and the authors thought that this could be due to the high dielectric constant of supercritical water, which would support heterolytic reactions and/or direct attack of water as a nucleophile. Infrared (IR) spectroscopy data showed greater carbonyl functionality in the case of the supercritical water depolymerised product, supporting the idea of direct attack by water as a nucleophile. The authors also thought that this functionalisation of the product may be a key feature in adding value to the recycled waste material.

The use of organic solvents to facilitate the devulcanisation of rubber without the addition of any other chemical agents has been explored by some workers, and the results published in patents. For example, Hunt and co-workers have published a US patent (US 5,891,926) that involves the devulcanisation of rubber at a temperature between 150 and 300 °C, at a pressure of a least 3.4×10^6 Pa, in the presence of the solvent 2-butanol. After the devulcanisation has occurred the rubber is present as a 'slurry' in 2-butanol. A second US patent published by Kovalak and Hunt (US 6,548,560) is similar to the first, but it cites the use of both alcohols and ketones, and has a critical temperature within the approximate range of 200–350 °C. Both processes are batch processes, which require relatively long periods of time to be effective.

A lot of relatively recent work, using specially prepared sulfur-cured NR and polyisoprene compounds, has been carried out by a group at Kyoto University led by K. Kojima [10]. In addition to studying the fundamental factors surrounding the use of supercritical CO_2 as a reaction medium with polyisoprene-type rubbers, they have used supercritical CO_2 in conjunction with a number of devulcanisation agents, for example:

- Diphenyl disulfide (DPDS)
- Thiophenol/*n*-butylamine mixture
- Triphenyl phosphine
- Mercaptobenzothiazole

An autoclave was used for the devulcanisation reactions and, in addition to the performance of the different agents, the influence of the CO_2 pressure and process temperature on the reaction were also investigated [11]. They found that the thio/amine and DPDS agents were the most effective at devulcanising the rubbers, with a soluble fraction of 100% (unfilled compounds) being obtained, but that this product had a relatively low Mw (number average molecular weight (Mn) ≈ 30,000 Da). They also carried out work (using only DPDS) to determine the effects that different levels of carbon black have on the devulcanisation reaction [12]. Regardless of the amount of carbon black, soluble fractions of between 20 and 40% were obtained. A number of other properties (e.g., G' and tan δ values) were also obtained, using dynamic mechanical analysis (DMA), and these confirmed that carbon black did not interfere with the devulcanisation reaction. In addition to model compounds, a tyre truck compound was also devulcanised. New compounds containing varying amounts of devulcanised rubber (DR) were produced, and the results obtained showed a 10% reduction in tensile strength for a level up to 40 phr of DR.

Verbruggen and co-workers [13] prepared a model NR gumstock (i.e., cure system only, no fillers or other additives) compound and subjected a thin film (300 μm thick) of it to DPDS at 200 °C. Solvent extraction (with acetone and tetrahydrofuran), swelling experiments (in toluene) and chemical analysis by GPC and nuclear magnetic resonance (NMR) spectroscopy were used to assess the degree of devulcanisation and to characterise the samples produced. It was found that complete network breakdown was obtained with 2.4% w/w of DPDS after 2 h of heating, but that both crosslink scission and main-chain scission had occurred.

4.4 Mechanical devulcanisation processes

4.4.1 Background to process

The processes that are covered by this section are those where the rubber is subjected to high shear forces and sometimes temperatures that are above ambient, but there is no use of any chemical agents. The lack of chemical agents can be regarded as a distinct advantage of this type of process in that their absence can yield benefits and advantages in a number of areas, for example:
- Cost
- Reduced odour
- Use with non-sulfur-cured rubbers
- Health and safety
- Environmental

In some cases (e.g., the Watson Brown high stress mixing (HSM) process) specialist equipment has been designed and manufactured that subjects the rubber to high levels of shear/stress under carefully controlled conditions. In other cases, existing commercial processing equipment (used for either rubber or plastics) has been employed, sometimes with a degree of modification, to carry out the devulcanisation.

As with the other devulcanisation process categories, the type of equipment that is used will have a strong bearing on whether the process is a continuous process or a batch process. A continuous process can be set up using an extruder and, in order to generate the high levels of shear that benefit this process, a twin-screw extruder is usually used with the elements along the screws set to a high shear configuration. A batch process can be set up using either a two-roll mill, an intermix or specially designed equipment.

By the generation of high shear forces within the rubber, these processes mainly work by exploiting the difference in the 'elasticity' of sulfur–sulfur bonds in the crosslinks compared to the carbon–carbon bonds in the main-chain polymer molecules. However, as the Watson Brown HSM process has demonstrated (Section 4.4.2), this type of devulcanisation process does not have to be confined to sulfur-cured rubbers. Irrespective of the type of cure system that has been used (i.e., sulfur-type, peroxide, metal oxide and so on), there is often some difference in the intrinsic chemical properties, particularly the degree of flexibility, associated with the chemical bonds present in the crosslinks compared to the bonds in the polymer main chain. The high shear conditions that are used in these types of processes can exploit these differences to preferentially break the crosslinks.

In common with the thermal processes (Section 4.3), these processes and mechanical processes that use chemical agents (Section 4.4) often use supercritical fluids (e.g., supercritical CO_2) as a process aid. A supercritical fluid is chosen that is compatible with the rubber that is being devulcanised so that it swells the rubber within the process equipment and so facilitates devulcanisation by ensuring a high degree of fill and, hence, shearing efficiency. It is, however, harder to maintain the fluid in a supercritical state in these types of processes, due to leakage and loss of pressure, than in sealed vessels (e.g., autoclaves), as are often used in thermal processes.

Due to the advantages of this type of process, in addition to the large amount of research that has been undertaken, a number of companies have established commercial operations, and a selection of these are reviewed below, together with examples of the research that has been conducted.

4.4.2 Examples of processes and published studies

Zhang, Lu and Liang of the University of Sichuan in China [14] have devulcanised NR at ambient temperatures, without the use of any chemical agents, using a self-designed pan mill-type mechanochemical reactor capable of exerting relatively strong shear forces on the rubber. The results obtained showed a large reduction in the gel fraction of the NR sample, revealing that it had been substantially devulcanised by the process, and the insignificant degree of reduction in the Mw of the sol fraction (i.e., the soluble fraction) indicated that the degree of degradation of the backbone of the material was low. The number of milling cycles was shown to have a significant effect on the mechanical properties of the NR, with the tensile strength being increased to around 7 MPa after 30 cycles, which is 60% of the strength of the virgin NR. Products that had very high (up to 973%) elongations at break could be obtained, which is equivalent to virgin vulcanisates. The workers concluded that solid-state mechanochemical milling is a simple, low-cost method for the devulcanisation of rubber at ambient temperature without the use of any chemicals.

Maridass and Gupta [15] have investigated the application of a laboratory-scale, intermeshing counter-rotating twin-screw extruder for devulcanising ground tyre rubber (GTR). One design used screws that had constant root and flight diameters and a length-to-diameter ratio of 10, while another had a progressively increasing root diameter. Although a range of extruder temperatures were evaluated (100–180 °C), the best results were obtained when a relatively high temperature (170 °C) was used with a screw speed of 30 rpm. The DR was mixed with 30 phr of NR and high-abrasion furnace-type carbon black in various proportions (15 phr of black proving to be the optimum level, with a tensile strength of around 16 MPa being obtained). A cure system based on cyclohexyl benzothiazole (CBS) and sulfur was also incorporated into the DR prior to curing.

Balasubramanian [16] has described a devulcanisation process that uses a counter-rotating twin-screw extruder to devulcanise GTR. The DR was then blended with virgin NR in various proportions and the blends revulcanised using a sulfur cure system. The Mooney viscosity, cure characteristics and mechanical properties of the resulting vulcanisates were characterised and a four-parameter rheometric equation, based on the standard logistical model for the curing behaviour of extrusion processed blends, was derived and validated for the different levels of virgin NR.

A group at Jiangsu Polytechnic University [17, 18] have used a twin-screw extruder to continuously devulcanise scrap tyre rubber and have studied the influence that screw speed and process temperature have on the degree of devulcanisation and the physical properties of the material once it had been revulcanised. A scanning electron microscope (SEM) was also used to study the morphology of the DR and the interfacial bonding that occurred within the revulcanised products.

Another Chinese group at Nanjing University of Technology have also used an extrusion process to devulcanise waste tyre rubber [19]. In their process, a thermoplastic material (polypropylene, PP) was added to the extruder during the devulcanisation process to ultimately create a toughened thermoplastic product. The devulcanisation stage was characterised in terms of the degree of devulcanisation that was achieved, and the toughened PP in regard to its physical and chemical properties, such as double-bond content and impact strength.

Macsiniuc and co-workers [20] from two research institutes in Canada have used both internal mixer and twin-screw extrusion processes to devulcanise waste ethylene-propylene-diene monomer (EPDM) rubber. The degree of devulcanisation achieved in products from both systems was characterised using a modified version of the American Society for Testing and Materials D-6814 method for determining crosslink density by swelling specimens in a solvent, and the particle morphologies were examined using a SEM. The results they obtained showed that a reduction in crosslink density could be obtained with the use of a chemical devulcanising agent, but that it was also possible to increase the crosslink density under certain operating conditions. Physical models were proposed by the workers in the paper to account for the observed changes in crosslink density.

Waste butyl rubber has been both devulcanised and re-revulcanised by workers in the East China University of Technology [21] under conditions of high shear and high temperature in a Haake Rheometer. They investigated the effect of the process temperature on the degree of devulcanisation and also compared the damping and physical properties of the waste rubber, DR and re-cured materials.

One devulcanisation process that has received a lot of publicity over the past few years is the Watson Brown HSM process, and a review of its benefits and applications was provided by Brown [22] to a meeting of the Institute of Materials in London in 2008. The HSM technology involves subjecting the waste rubber to a high shear environment within a specially designed mixer (under controlled conditions, i.e., temperature, time and gap size) to break the sulfur crosslinks, adding an optimised amount of sulfur to the resulting compound on a two-roll mill and then re-curing it. Figure 4.1 shows images of the stator and rotor in an HSM and a photograph of an HSM with a capacity of 20 litres is shown in Figure 4.2. In the presentation, some data was shown where the HSM technology had been used to recycle a number of post-consumer products, for example a nitrile sealing material, an EPDM door seal, and different waste tyre tread compounds. In the case of the waste tyre treads, they had then been mixed with some sulfur, cured and then tested. The results showed that, in one case, increasing the amount of sulfur added at the two-roll mill stage increased the tensile strength from 10.2 MPa to 14.3 MPa, with a corresponding increase in elongation from 161% to 236%, and an improvement in the surface finish of the product. Other recycled tyre tread data presented showed that tensile strengths and elongations at breaks of over 18 MPa and 470%, respectively, were possible. In addition to applying the technology to tyre manufacturing, its potential for use in the production of roads, surface coverings, and flooring and roofing products was also described. An overview of the Watson Brown HSM Ltd operations in Europe and North America, the performance of the HSM process, and the properties of products in which HSM DR had been incorporated are described in a paper presented by Maddever and Watson [23] to the *Fall 2010 Meeting of the American Chemical Society (ACS) Rubber Division*

Figure 4.1: Stator/rotor parts of a Watson Brown HSM. Reproduced with permission from the Institute of Materials, Minerals and Mining, London, UK. ©2008, Watson Brown HSM Ltd.

Conference and a paper presented by Brown and co-workers to the *ACS Rubber Division Fall Meeting* in 2009 [24]. In 2010 it was reported [25] that a production facility near Berlin in Germany that used the HSM process had been set up to devulcanise rubber that was supplied by rubber companies for incorporation back into their own products. One advantage that the process is claimed to have over a number of others is its capability to devulcanise both conventional sulfur-cured diene rubbers (e.g., styrene-butadiene rubber (SBR) and NR) as well as higher-value elastomers, such as hydrogenated nitrile rubber and fluorocarbon rubber, which are cured using peroxides and other types of cure system. The process is also claimed to be economically viable for mid-priced elastomers such as acrylics, chlorosulfonated polyethylene rubber, silicone and butyl-based rubbers. Tests carried out on the devulcanised products have shown that their viscoelastic properties are almost identical to the uncured

Figure 4.2: Watson Brown 20 litre HSM. Reproduced with permission from the Institute of Materials, Minerals and Mining, London, UK. ©2008, Watson Brown HSM Ltd.

virgin elastomers, which enables them to be used at levels of up to 30% in new compounds without prejudicing their processing properties. Some other technical and commercial information about the German plant was reported in an article in the *European Rubber Journal* in 2009 [26].

Meysami and Tzoganakis [27, 28] of Waterloo University in Canada have used an industrial-scale twin-screw extruder with supercritical CO_2 to devulcanise waste tyre rubber, tyre tread rubber and EPDM door seal rubber. Their approach uses the swelling effect that the CO_2 has on the rubber and the high shear forces generated within the extruder to preferentially break the sulfur–sulfur crosslinks. The workers claim that a reasonably high output of product is obtained (e.g., 60 kg/min using a 50 mm twin-screw extruder) and they have studied the influence that various process parameters (e.g., screw speed and screw configuration) had on the degree of devulcanisation. Various amounts of the devulcanised products were blended with virgin rubber to produce vulcanisates, and the cure performance and mechanical properties of these showed that good- quality products had been obtained. The possibilities of setting up a larger, pilot-scale plant have been considered by the group.

Shen and co-workers [29] used a twin-screw extruder to devulcanise waste tyre rubber and then blended the resulting DR into high-density polyethylene (HDPE) to produce thermoplastic elastomers (TPE). They found that the high shear conditions during the extrusion process induced chain scission and oxidative degradation of the GTR. The reduction in crosslink density was apparent in a decrease in the gel content of the rubber/HDPE blends and an increase in the melt flow of the materials.

During the manufacture of medical syringe stoppers, crosslinked rubber waste can be generated (e.g., in the form of moulding 'flash' or reject mouldings) and Maridass and Gupta [30] have used a technique known as response surface methodology to predict the optimal processing parameters (e.g., barrel temperature and screw speed) to devulcanise the powdered waste in a twin extruder. A central composite rotatable design for two variables at five levels was chosen as the experimental design, and the results obtained after measuring the responses were fitted as two-variable second-order equations. The contour plot, response surface plot, Pareto chart and main effect plot were generated, and from these the optimum properties were predicted as being a temperature of 180 °C and a screw speed of 27 rpm. Experiments were then performed to devulcanise the stopper waste and these values were confirmed as being the optimum for these two parameters.

Kang and co-workers [31] have reported on the effects of processing conditions on the reversion that takes place when vulcanised NR is devulcanised using a single-screw extruder. To monitor the levels of reversion they used a range of analytical techniques including Fourier-transform infrared (FTIR) spectroscopy, thermogravimetric analysis (TGA), rheological, and sol fraction determinations and put the results that were obtained into the context of the development of effective continuous devulcanisation processes.

An Iranian team [32] have devulcanised tyre tread using a twin-screw extruder and determined how varying the barrel temperature and the screw speed of the extruder affected the degree of devulcanisation. The results showed that percentage devulcanisation was mainly dependent upon screw speed, whereas the barrel temperature had a greater effect on the sol fraction that was obtained from the rubber. Once the rubber had been devulcanised, the team added curing agents to it and revulcanised it. They found that it behaved upon re-curing in a similar way to conventional rubbers. Blends of the DR with virgin rubber (ratio of 15:85) were produced and cured successfully. The mechanical properties, compression set, hardness and resilience of this blend were determined and the results showed that they were slightly inferior to the 100% virgin compound.

Tzoganakis and Zhang [33] of the University of Waterloo in Canada have developed a CO_2/twin-screw extrusion process that is reported to be capable of devulcanising rubber crumb from waste tyres. The primary purpose of the CO_2 was to act as a processing aid and enable the devulcanisation to be carried out using the mechanical shear and extension forces produced by the extruder. The affect that varying a number of parameters (e.g., screw configuration, screw speed, mesh size of crumb, feed rate of crumb, process temperature and amount of CO_2) had on the devulcanisation reaction was investigated. The configuration of the screw, its speed and the process temperature were found to have the greatest influence. Soluble fractions of up to 60% were obtained, with chemical analysis data showing that additional double bonds had been created within the polymer molecules, and that chain scission has occurred due to the breakage of carbon–carbon bonds, in addition to breakage of the crosslinks.

A research group in Canada [34] have used supercritical CO_2 in a twin-screw extruder to devulcanise waste rubber from the automotive sector. They have used the process on both waste tyre rubber and EPDM-based products, such as door seals, and claim to have achieved good results with both types of material. The process uses that fact that as the CO_2 swells the rubber in the high shear environment within the twin-screw extruder, the sulfur–sulfur crosslink bonds break preferentially compared to the carbon–carbon main-chain bonds (Section 4.2).

A high-shear, mechanical devulcanisation process that uses no chemicals or additives has been developed by H S Polymer Reprocessing in the UK and was the subject of a paper presented by Savage and Hartley at a *High Performance Elastomers Conference* [35]. The HSP process is capable of devulcanising high-value elastomer compounds that have been vulcanised using both sulfur and nonsulfur-curing systems. It is claimed that any changes from the properties of the virgin material are predictable and usually within an acceptable level at percentage values of 20% and below. Examples of nonsulfur-cured rubbers that have been successfully reclaimed in this way are fluorocarbon, hydrogenated nitrile, and fluorosilicone.

4.5 Mechanical devulcanisation in the presence of chemical agents

4.5.1 Background to process

The processes that are included in this section are those that employ both shear forces and chemical agents to devulcanise the rubber. In a number of cases, heat is also deliberately applied to the rubber by the use of a heated extruder barrel or mill. Even in those cases where a process may be described as operating at 'ambient', they will normally function above typical ambient temperatures due to the frictional heat generated.

A number of chemicals have been used to devulcanise rubber and a very good review of the different types that are used to devulcanise and reclaim rubber has been published recently by Myhre and co-workers [36]. In this review they divided the types of organic chemical agents that could be used into the following categories:

- Those that react with free radicals to prevent reformation of the crosslinks, e.g., disulfides, thiols and phenolic compounds
- Those that cleave the sulfur crosslinks by a nucleophilic mechanism, e.g., amines and their derivatives

Also, in addition to referring to the work carried out by Tobolsky [6] and Murakami and Ono [7] (Section 4.2), a lot of the devulcanisation work that has targeted the chemical nature of the sulfur bonds in rubber has used the findings of the 'chemical probe' research carried out by workers at TARRC (previously called NRPRA and MRPRA) in Brickendonbury, Hertforshire, UK [37], and many others [38, 39].

When these targeted chemical reactions are carried out in a static environment (e.g., an autoclave) only heat and a swelling agent to open up the rubber matrix and allow good penetration into the rubber matrix are required. These types of system have already been reviewed in Section 4.3. These batch experiments are usually only carried out on a small scale though, and it is more usual for mechanical work, and possibly some heat, to be used in conjunction with the chemical agent if a larger scale, more commercially viable process (e.g., in an intermixer or an extruder), is being developed.

Some examples of the chemical agents that have received the most attention in these types of processes are listed below:
- Diphenyl disulfide
- Diallyl disulfide
- Amine/thiol mixtures
- Standard rubber industry accelerators (e.g., CBS, diphenyl guanidine, mercaptobenzothiazole disulfide (MBTS) and zinc diethyl carbamate; these can be used with or without other rubber chemicals, such as peptisers)

Some of these chemical agents (e.g., the amine/thiol mixtures) were used in the 'chemical probe' rubber chemistry research work being undertaken at TARRC in the 1950s and 1960s to establish the chemical nature and processes involved during the sulfur vulcanisation of diene rubbers, such as NR.

Some of the devulcanisation agents that have been used can have disadvantages, for example:
- They can impart an unpleasant odour to the DR.
- They can require relatively high temperatures to function, which is more demanding in terms of energy consumption, and can cause the rubber (particularly diene-type rubbers) to degrade.
- They can have health and safety concerns associated with them.
- They can lead to an increase in factory emissions of volatile organic hydrocarbons.
- They can be relatively expensive.
- They can be less than 100% efficient in their targeting of the sulfur crosslinks.

The use of CO_2 in the supercritical, or liquid state, to facilitate the devulcanisation of rubber has already been mentioned in Sections 4.3 and 4.4. However, because it was used on its own as a non-reactive 'process aid' in those thermal and mechanical processes, it was not regarded as acting as a chemical devulcanisation agent, and so those systems were not included in this section. CO_2 has also been used in these types of processes by some workers to swell the rubber and, depending of their solubility, to also solubilise the devulcanisation chemicals. These mechanisms help to increase the penetration, at the molecular level, of the chemical agents into the rubber matrix and so offer the potential of improving their effectiveness. Retention of the CO_2 in the system, as it will rapidly try and revert to a gas at the process temperatures being used, can be more of a problem with certain equipment designs than others. For example, retaining the CO_2 when using a twin-screw extruder that has a 'clam shell' design is harder than keeping it within a single-screw extruder that has a sealed barrel. In addition to its low cost and low toxicity, the use of CO_2 can also bring other advantages. By swelling the rubber, it can help to ensure that the barrel is fuller and so increase the amount of work (i.e., shear) that is carried out on the rubber. As stated in Section 4.2, shearing the rubber is one of the mechanisms that can be used to break the sulfur–sulfur crosslinks. Also, as it volatilises from the rubber at the end of the process, it can act as a cooling agent to bring the temperature down and reduce the possibility that it will degrade.

4.5.2 Examples of processes and published studies

As mentioned above, one of the principal problems that can be encountered with these types of systems is when they are applied to waste diene-rubber-based products, such as waste tyres, as the high temperatures that are required to enable some of the chemical agents to react with the sulfur crosslinks (e.g., the diphenyl sulfides, which usually require temperatures of around 150 °C or more) also leads to degradation (e.g., reversion and/or oxidation) of the rubber with a reduction in Mw and, hence, important physical properties such a tensile strength.

The Israeli company Levgum have patented a process (US 6,831,109) based on a chemical method using a combination of two types of chemical compound: a urea-type compound and a dicarboxylic acid-type compound. The former compound dissociates to produce an organic cation and an amine, and the latter as a promoter of this dissociation and an acceptor of the amine. The ionic compound that results from this reaction is claimed to be more selective towards crosslink destruction than the active species generated by other chemical agents (e.g., by rubber accelerators) and achieves a more efficient penetration into the rubber matrix. Levgum have referred to this devulcanisation agent as the 'Ecological De-Vulcaniser' and have said that it is derived from the agricultural industry and is used at a relatively low level (3%) [40].

The Levgum process is designed to function at ambient temperatures using multiple passes (around 20) through a two-roll mill and can be used to devulcanise waste rubber originating from a number of sources (e.g., tyres, conveyor belts and so on) into a high-quality product. An article in the *European Rubber Journal* explained that the business model is based around licensing the technology to partners, each with an exclusive regional licence [41]. In the same article, it is stated that Levgum claim that their process does not use harmful chemicals, does not produce or emit any hazardous waste or gases, and is 100% environmentally safe. By 2011, the number of licensees had grown to six franchises in five continents.

The DeLink process, which was originally developed in the mid-1990s, addresses the relatively poor thermal stability of diene rubbers by employing a chemical-based system that is capable of devulcanising rubber at relatively low temperatures (e.g., <70 °C). The process is based around a masterbatch, which is added to the vulcanised rubber using either a two-roll mill or intermix at relatively low mixing temperatures. This masterbatch consists of a blend of conventional rubber chemicals, e.g., accelerators, zinc oxide and stearic acid. The DeLink system is regarded to have an advantage in that the rubber does not require re-compounding as the residual chemicals within it enable it to be revulcanised at elevated, standard curing temperatures (e.g., 150 °C). One of the advantages claimed for this process is that it reduces the total heat history that the rubber experiences, and hence the amount of thermal degradation that takes place. At present, new rubber products that have been produced using the DeLink technology can be found marketed under the trade name 'Green Rubber'. An example of its application is in the soles of the 'Earth Keeper' range of footwear marketed by Timberland.

One of the more recent presentations describing the technical aspects of the DeLink process, its economic and environmental benefits, and the properties of rubber compounds containing the DR was presented by Nieto of Green Rubber Global at the *ACS Rubber Division Fall Meeting* in 2009 [42]. It is also the case that DR that results from the use of the DeLink technology can be referred to as De-Vulc [43].

A devulcanisation process that has some similarities to the DeLink process has been published as a US patent by Tang (US 6,590,042). In this process, a specially designed twin-screw extruder is used, with the waste rubber in a crumb form, and a 'reclaiming agent' that includes the following rubber compounds:
- *Tert*-butyl-2-benzothiazole sulfenamide
- Zinc mercaptobenzothiazole (ZMBT)
- Mercaptobenzothiazole (MBT)
- Tetramethyl thiuram monosulfide
- Zinc oxide
- Stearic acid
- Zinc salts of fatty acids
- Sulfur

Other combinations of ingredients, such as one sulfenamide, two mercapto group (i.e., ZMBT, MBT or MBTS) accelerators, a thiuram accelerator, zinc salts of fatty acids, and one activator (e.g., stearic acid) are also covered by the patent.

The process is claimed to work with all sulfur-cured rubbers and, as it is an extruder-based system, is can be operated continuously with an output of up to 750 kg/h.

The fact that there are potential disadvantages with some of the chemical agents that are used to devulcanise rubber has presented opportunities for manufacturers to develop new chemical devulcanisation agents that address them. An example of these novel chemical devulcanisation agents, DB-26, has been developed by Polymeright. The technology that is associated with the use of this chemical agent, and the benefits that it imparts when used, were presented by the company at an ACS meeting in Pittsburgh in 2009 [44].

The use of supercritical CO_2 to facilitate the devulcanisation of rubber with chemical agents has been investigated by a number of workers. As mentioned already, this compound can assist the process in a number of ways, for example by acting as a swelling agent to open up the rubber matrix and so act as a carrier for the chemicals, aiding their penetration and, hence, their interaction with the crosslinks.

A considerable amount of research using a system employing CO_2 was carried out from 2006 until 2012 by a consortium led by Smithers Rapra [45]. This consortium was responsible for two projects, DevulCO_2, funded by the UK's Department of Trade and Industry, and ReMould, funded by the UK's Technology Strategy Board. The first project, DevulCO_2, ran from November 2006 to April 2009, had a budget of around £1 million and had six UK companies in the project consortium. The objective of this project was to develop a novel process to devulcanise waste tyre rubber for manufacturing high-quality products and its target markets were the GRG sector, e.g., damping pads, instrument boots, gaskets, bushes, seals and so on. This project did not target the manufacture of new tyres; this type of work was carried out within its successor project, the ReMould project (reviewed below).

The technology developed during the DevulCO_2 project was a success and a UK patent describing it was published on 21st December 2011 (GB 2,481,205). The process used rubber crumb (20–60 mesh) from truck tyres as its raw material and the devulcanisation step involved the use of extruders (either twin-screw or single-screw types) at a relatively low temperature and was assisted by the incorporation of CO_2. A wide range of commercial products were successfully produced using the DevulCO_2 devulcanised tyre rubber by members of the consortium (some examples are shown in Figure 4.3) and the conclusions from the results obtained were that it could be substituted for virgin rubber in a number of cases, but that there were potential problems when good tear strength was required or where the specification was very high.

The second research project, ReMould, ran from October 2009 until June 2012 and had a budget of about £1 million. Its aims were to build on the achievements of the DevulCO_2 project by carrying out the following:

Figure 4.3: Selection of compression and transfer moulded products made from DevulCO_2 rubber. Reproduced with permission from the DevulCO_2 Project Consortium, Smithers Rapra, Shawbury, UK (Lead Partner). ©2012, DevulCO_2 Research Project Consortium.

- Commercial scale-up of DevulCO_2 devulcanisation technology.
- Transfer of DevulCO_2 devulcanisation technology to sulfur-cured GRG products, e.g., nitrile rubber matting and EPDM automotive weather strip.
- Improve processibility to enable extrusion profile work, e.g., manufacture of recycled weather strip.
- Retreading of tyres using relatively high levels (e.g., >25%) of DevulCO_2 rubber.

The scale-up goal was achieved and the DevulCO_2 devulcanisation technology was successfully transferred to a number of sulfur-cured rubbers, including nitrile rubber, EPDM rubber and NR. In addition, the results of the evaluation of the performance of the DevulCO_2 rubber in tyre retread compounds were encouraging. A more extensive description of these two projects and the results of the manufacturing trials that were undertaken during them, is provided in Chapter 5, Section 5.2.2.1.

A paper published by Antonio [46] from Rubber Resources BV indicates that combining a particular type of scrap rubber with the right process conditions (temperature, amount of frictional heating/shear, addition of a small amount of water, use of a devulcanisation agent and so on) can result in more effective devulcanisation and a revulcanised product with better physical properties. Some of the results that have been published by Rubber Resources BV using their system are shown in Chapter 5, Section 5.2.2.4. Two recent photographs showing examples of the production processes that are carried out by Rubber Resources BV are provided in Figures 4.4 and 4.5. In Figure 4.4 the extrusion head for the production of Ecorr® natural rubber reclaim (RNR) is shown, and in Figure 4.5 the refiner device for the production of Ecorr® butyl rubber reclaim (RBR) is shown.

Figure 4.4: Extrusion head for the production of Ecorr® RNR. Reproduced with permission from Rubber Resources BV., Maastricht, The Netherlands. ©2014, Rubber Resources BV.

Figure 4.5: Refiner device for the production of Ecorr® RBR. Reproduced with permission from Rubber Resources BV, Maastricht, The Netherlands. ©2014, Rubber Resources BV.

Diallyl disulfide is another disulfide compound that has received attention in the literature. Jana and Das [47, 48] subjected used tyre tread samples to a novel, in-house-developed, open two-roll cracker-cum mixing mill. As the name suggests, this mill had both a cracker zone for grinding the rubber and a mixing zone that mixed the rubber with the diallyl disulfide and some aromatic process oil at 110 °C. The DR produced by this process was revulcanised after the addition of sulfur, CBS and a retarder (thiophthalimide type), having a tensile strength of 17.6 MPa (original rubber 24.3 MPa) and an elongation at break of 590% (original rubber 596%). In addition to a good retention of physical properties, an advantage claimed for this process is that it is easier to re-compound curatives into the fine elastic powder produced by the method than rubber powder produced by cryogenic grinding.

A company called UAB Devulco and the Institute of Chemistry, both based in Vilnius in Lithuania, have developed a technology that uses a combination of mechanical shearing and chemicals, without any additional heating, to produce what is referred to as 'fine reclaim powder' (FRP) [49]. An eco-friendly devulcanising agent (in the range 2–5%) is used in the process, which can be carried out using conventional equipment (e.g., an extruder), and the agglomeration of the devulcanised powder is prevented tribologically to ensure high yields. The residual devulcanisation agent can be removed by drying. A high level of devulcanisation is claimed (>90%), along with conversion rates of 95%, for a range of waste rubber types. The results obtained in a number of trials undertaken using the 20–40 mesh, 40–60 mesh and 60–80 mesh FRP in blends with a virgin NR compound at levels of 20 and 45 phr showed that the FRP had better physical properties and was easier to process compared to equivalent compounds using 20–40 mesh truck tread buffings and conventional, sheeted reclaimed rubber.

The use of the amine hexadecylamine (HDA) to devulcanise sulfur-cured EPDM rubbers has been described by Dijkuis and co-workers [50]. Conventionally and efficiently vulcanised (EV) EPDM rubbers were devulcanised in a Brabender Plasticorder with HDA under various conditions. The effects of temperature, time, shear and amount of reclaiming agent on devulcanisation were examined using a design of experiments approach, and the devulcanisates were characterised by Mooney viscosity measurements, solvent extraction, crosslink density determinations and light microscopy.

The same devulcanising agent has also been used, along with shear force and heat, to devulcanise sulfur-cured EPDM by Sutanto and co-workers [51]. The group carried out the work using both a batch process, using a Brabender batch mixer, and a continuous one using an extruder. The results of the work carried out using the Brabender enabled them to propose an engineering kinetic model for the devulcanisation of EPDM. This model was derived by using rate-determining parameters, such as temperature, shear rate, reactant concentration and time, and it covers all the phenomena happening during the devulcanisation process, without distinguishing them on a molecular level. The amount of unreacted HDA in the EPDM after each devulcanisation experiment was quantified by extracting the rubber with hexane, and analysing the extract using a gas chromatography-based method. The workers believe that although their model is not an elementary one, it is able to predict the decrease of the reactants adequately so the results can provide an idea about the reaction profiles and will be useful for designing a reactor and contribute to further research into the kinetics of devulcanisation. The results obtained using HDA at a concentration of 3×10^{-4} mol HDA/g rubber (the optimum concentration derived from an initial screening study) showed that the rate of side reactions involving HDA is much lower than the reaction rate between HDA and the crosslinks, which is the desired situation. The workers believed that it may be possible for the structure of their model, and the modelling procedure, to be applied to other devulcanisation processes where both shearing action and devulcanisation agent are used.

A team of workers in India [52] have described the use of tetramethyl thiuram disulfide (TMTD) in the presence of Spindle oil for the devulcanisation of ground tyre rubber at near ambient temperatures. Small-scale (~100 g batches) devulcanisation experiments were carried out on a two-roll mill and the products produced characterised using a variety of analytical techniques (e.g., dielectric analysis and TGA). The amounts of TMTD and Spindle oil were 2.75 and 10%, respectively. The DR was mixed with virgin, uncompounded NR and a cure system, with the DR content varying from 0 to 60%. Vulcanisates were produced, which were characterised by both physical (e.g., tensile strength and swelling) and analytical (e.g., SEM and DMA) tests. The tensile strengths that resulted were found to vary from 7.7 MPa (60% DR) to 14 MPa (0% DR).

A group of workers at Twente and Groningen Universities in the Netherlands [53] have carried out devulcanisation experiments on waste EPDM automotive weather strip in a co-rotating twin-screw extruder with HDA as the devulcanisation agent. Once it had been devulcanised, the EPDM was blended with different proportions of virgin EPDM roof sheeting compound masterbatch and vulcanised. The products obtained from these blending experiments were extensively tested to determine a range of characteristics, including degree of dispersion of the devulcanised EPDM, physical properties, compression set performance and weathering resistance. A group at Twente University in the Netherlands [54] have also investigated the mechanism involved in the devulcanisation of two different EPDM compounds using HDA as the devulcanisation agent.

In addition to the work referred to above, workers at Twente University [55] have also carried out an investigation on the effectiveness of three different devulcanisation agents on the devulcanisation of two different types of EPDM rubber. The three chemical agents were:

- HDA
- Diphenyl disulfide
- *o,o′*-Dibenzamidodiphenyl disulfide

The two different EPDM compounds were one that had been cured using a conventional sulfur cure system and so had a relatively high proportion of polysulfidic crosslinks, and one that had been cured using an EV-type sulfur cure system and had a relatively high proportion of mono-sulfidic crosslinks. The devulcanised and revulcanised products that resulted from these combinations were characterised and then the efficiency of devulcanisation correlated with the suitability of the devulcanised EPDM for revulcanisation. This approach also enabled the researchers to ascertain the influence that the three different chemical agents had on the physical and mechanical properties of blends of the DR with virgin EPDM of the same composition. The results showed that hexadecyclamine was the most versatile devulcanisation agent and could be used to effectively devulcanise both conventional and EV cured EPDM compounds.

Another group at Twente University and DSM Research [56] studied the effects of different diene monomers on the devulcanisation properties of an EPDM rubber that had been cured by both sulfur and peroxide cure systems. The diene monomers included in the project were ethylidene norbornene (ENB), dicyclopentadiene (DCPD) and hexadiene (HD), and their work also investigated the influence that the presence or absence of devulcanising agents had on the process. To measure the degree of devulcanisation that took place in the experiments, they used equilibrium swelling in solvent, and also determined the sol fraction of the samples. The results showed that ENB rubbers devulcanised by random scission for peroxide-cured compounds and by crosslink scission in sulfur-cured compounds. HD rubbers were found to devulcanise by crosslink scission for peroxide-cured compounds and by both crosslink and main-chain scission in sulfur-cured rubbers. The DCPD rubbers did not devulcanise, but showed an increase in the level of crosslinks present instead. An assessment of the devulcanisation mechanisms that existed in the presence and absence of devulcanisation agents was also undertaken during the programme of work.

Disulfide compounds have been known to behave as devulcanisation agents for sulfur-cured rubbers for a number of years. One of these compounds, DPDS, has been a particularly popular choice of devulcanising agent. Workers have found that relatively high (e.g., 180 °C) temperatures are required for it to function as a devulcanising agent and this can cause problems with thermally sensitive rubbers such as NR and SBR. Also, at lower temperatures (e.g., 140 °C) it has been found to exhibit properties that are similar to that of a vulcanising agent.

Roy and co-workers [57] at the Indian Institute of Technology have evaluated the performance and the chemical behaviour of a range of disulfide compounds when applied to waste tyre rubber. The DR that were obtained in each case were revulcanised to assess their cure characteristics, and then samples tested to determine their mechanical properties, morphology, dynamic properties and thermal stability.

A group of Chinese workers [58] have carried out work to devulcanise a NR compound using supercritical CO_2 in conjunction with the chemical devulcanising agent diphenyl disulfide. During the study they investigated the effects of a number of parameters, such as CO_2 density, temperature, pressure and reaction time. They also used two analytical techniques, GPC and NMR, to study the products of the various devulcanisation experiments.

Workers at the Iranian Polymer and Petrochemical Institute [59] have utilised the Box–Behnken experimental design approach to study the devulcanisation of EPDM using a co-rotating twin-screw extruder with DPDS as a devulcanisation agent. A number of different processing conditions were evaluated and the products characterised by sol fraction, Mooney viscosity and mechanical properties. The data obtained by the experiments were analysed using response surface methodology and presented in terms of the effects of barrel temperature, screw speed and level of DPDS on sol fraction, Mooney viscosity, tensile strength, hardness and elongation at break.

Work has been carried out on waste butyl rubber at Qingdao University of Science and Technology by Xiankui and co-workers [60]. Waste butyl tyre curing bladders were taken and placed into a laboratory-scale synchronous rotor mixer along with paraffin oil and a devulcanisation agent of an undisclosed nature. Experiments were conducted to assess the impact that a number of processing parameters, such as degree of fill, mixing time and amounts of reagents, had on the quality of the product that resulted. Vulcanisates from the devulcanised butyl rubber were also prepared under a range of conditions to determine those that resulted in the optimum mechanical properties.

A commercial product called Polykrosh SK [61] has been produced by a Russian company by masticating scrap nitrile rubber with dibutyl phthalate, oil asphalt and a proprietary additive. This material has been incorporated into rubber compounds and the properties of these, and vulcanisates that have been produced from the compounds, have been investigated and reported.

4.6 Ultrasonic devulcanisation

4.6.1 Background to ultrasonic process

These types of processes make use of the energy supplied by ultrasonic waves to preferentially break sulfur–sulfur bonds within crosslinks. They offer advantages over some of the other techniques in that they do not normally use solvents, or chemical agents, which helps to control cost and means that some health and safety issues do not have to be taken into consideration. They also have the potential advantage of being able to be operated continuously through the use of extruders.

Organisations and companies using this technique usually employ extruders that interface into ultrasonic equipment, although it is possible to position the ultrasound device at a point along the extruder barrel. Rubber crumb particles are fed into the extruder, which heats, softens and, to a certain extent, consolidates them. This masticated rubber is then subjected to ultrasonic energy to affect the devulcanisation. Ideally, the system has been optimised so that the ultrasonic waves have the energy required to selectively break the sulfur crosslinks. However, in common with a number of the other processes, this one is not always entirely selective with respect to crosslink destruction and the energy imparted to the rubber matrix can also cause main-chain modification (e.g., additional intramolecular crosslinks), leading to an increase in glass transition temperature.

The technique has been found by workers such as Isayev to be suitable for use with a great variety of rubbery networks (Section 4.6.2), and in the presence of rigid filler particles. Whereas the ultrasound reactor and power settings to produce the desired devulcanisation performance vary significantly between the types of rubbers, the general nature of the network destruction and degradation are thought to be relatively similar. Post-process NMR studies on the different rubber systems have shown structural variations, and these are thought to be due to differences in the extent of backbone cyclisation and isomerisation. The fact that rubber that is devulcanised by ultrasound can differ in these ways from the original virgin material is not thought by its proponents to detract from its usefulness as an option for recycling and it may in fact be found to be an attribute for certain applications. Its relative efficiency in terms of energy usage is also claimed to be an advantage.

The exact mechanism by which the ultrasonic treatment causes devulcanisation is still under study. One theory is that acoustic cavitation occurs within the rubber and it is the collapse of these cavities that causes devulcanisation to occur. However, workers have also postulated that the collapse of these cavities was not the primary method of devulcanisation and that degradation of the network around the cavities should also be considered. The theories surrounding the mechanism by which devulcanisation is achieved by the use of ultrasound, and other important characteristics of the technique, are covered by a comprehensive review of ultrasonic devulcanisation written by Isayev and Ghose [62].

4.6.2 Examples of processes and published studies

Professor A. Isayev and his co-workers at the University of Akron have carried out work using this technique for over 25 years and they have published a very large number of patents, as well as publications in a variety of journals and conference proceedings. The Akron group have used both single-screw and twin-screw extruders in their work. A schematic diagram showing both types of extruder is available on Dr Isayev's website (http://www3.uakron.edu/isayev/) and this is reproduced in Figure 4.6. A photograph of a one of the ultrasonic extruders used at Akron is also available on this website and is reproduced in Figure 4.7.

A number of the earlier papers published by the Isayev team at the University of Akron in the 1990s sought to fully understand the process and how it functioned when it was applied to relatively simple rubbers (e.g., unfilled NR). They investigated the effects that the parameters and variables within the technology (e.g., amplitude of acoustic pressure and acoustic oscillations) have on its effectiveness to achieve devulcanisation [63], the mechanism that is at work [64–66], the effect of different types of reactors [67], and the properties (e.g., tensile strength) that are obtained from the revulcanised rubber [68].

Because of the interest in the recycling of waste tyre rubber, the University of Akron team has carried out a large number of studies on the application of ultrasound to the devulcanisation of SBR. These studies have evaluated the effect that variables such as crosslink type [69], styrene content [70], and crosslink density and molecular mobility [71] have on the results that are obtained.

For the same reason, its application to NR has also been studied, looking at the influence of differing levels of carbon black [72], and the properties of blends of devulcanised NR and virgin NR [73]. Some of the results of these investigations indicated that the devulcanisation of NR and SBR exhibited different properties. For

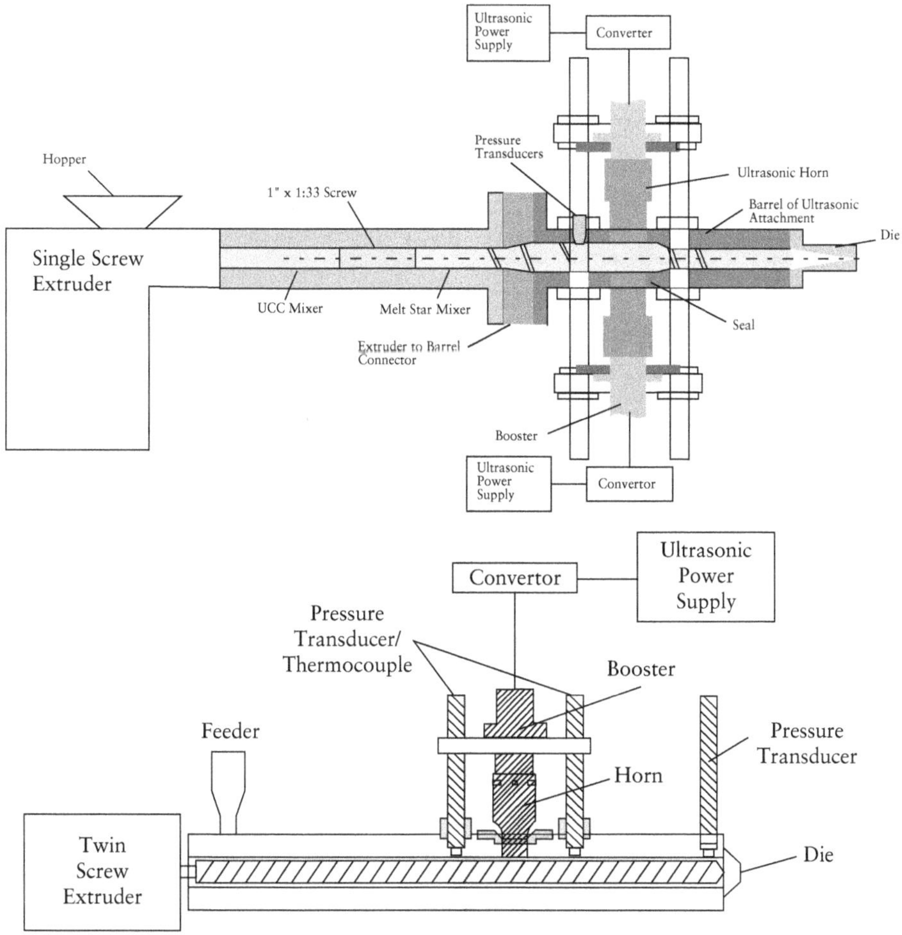

Figure 4.6: Schematic diagram of single-screw and twin-screw ultrasonic devulcanisation extruders. Reproduced with permission from Dr A.I. Isayev, The University of Akron, Akron, OH, USA. ©2014, Dr A.I. Isayev, The University of Akron.

example, the devulcanisation of NR was found to be independent of process parameters such as barrel temperature, die size, output rate, and amplitude of the ultrasound, but dependent upon the energy of the ultrasound, passing through a maximum at an intermediate energy level. With SBR, competition between the chain scission and crosslinking reactions was found. At low amplitude, crosslinking with gel formation was dominant, but at high amplitude, chain scission and branch formation were dominant.

Due to its origins, NR contains a number of non-rubber constituents such as proteins and it is very stereoregular, comprising 100% *cis*-1,4-polyisoprene. The synthetic versions of NR, which can be used for a number of applications (e.g., medical

Figure 4.7: Ultrasonic devulcanisation extruder. Reproduced with permission from Dr A.I. Isayev, The University of Akron, Akron, OH, USA. ©2014, Dr A.I. Isayev, The University of Akron.

products) usually have a *cis*-1,4 content of between 96 and 98% and, being produced by solution polymerisation, they are very low in non-rubber constituents. To establish if these differences influenced the devulcanisation behaviour of these closely related materials, Ximei and Isayev [74–76] have carried out a study on the application of continuous ultrasonic devulcanisation to a carbon-black-filled synthetic *cis*-1,4-polyisoprene rubber and a carbon-black-filled NR rubber. For these two systems, the influence that the carbon black had on the degree of devulcanisation, rheological properties and mechanical properties was investigated, as well as the impact that a processing oil and a cure retarder had on the devulcanisation and revulcanisation processes. One result was that the presence of oil was found to increase the amount of devulcanisation that took place, as shown by the increase in the ultrasonic power consumption and reduction in the gel fraction and crosslink density. The effect of ultrasonic amplitude on the energy consumption was also examined, and the results discussed in relation to the stereoregularity of the rubbers. For example, the observation with the carbon-black-filled IR was that the power consumption during the devulcanisation process increased with an increase in ultrasonic amplitude, indicating a greater extent of devulcanisation, which is a similar effect to that taking place with unfilled IR. The carbon-black-filled NR behaved differently, with the power consumption showing a maximum at intermediate amplitude, suggesting that less devulcanisation was taking place at higher amplitudes.

In addition to studies on NR and SBR rubbers, Isayev and Jeong Seok Oh have also carried out work on another of the diene rubbers that is often found in tyres, butadiene rubber (BR), which is often used in tyres as an anti-flex crack additive in the tread compound [77]. They applied a continuous ultrasonic devulcanisation system, which was fitted with a grooved barrel, to unfilled BR. The results obtained by a range of experiments in which a number of parameters were varied indicated that BR has a narrow devulcanisation window, outside of which a significant amount of degradation or no devulcanisation occurred. The analysis technique GPC was used to characterise the Mw of the sol fraction of both the devulcanised BR and virgin BR, and the results showed a significant Mw reduction and a broadening of the Mw distribution upon devulcanisation, indicating that devulcanisation and degradation were occurring simultaneously. Although the mechanical properties of the devulcanised BR were inferior to the virgin BR, thought to be due to breakage of the main chains, when the two were blended, a considerable enhancement was observed. For example, once the level of virgin BR had reached 50% in the blend, the values of the elongation at break were greater than those of the pure virgin material. The tensile strength values also increased beyond what would be predicted as the amount of virgin BR increased. These observations were thought to be due to the good bonding of ultrasonically devulcanised BR to virgin BR, a phenomenon that the workers had also observed with NR. The devulcanised BR displayed an inferior level of thermal stability compared to the virgin BR, probably due to the chemical changes that had gone on within the material as a result of the ultrasound treatment.

Butyl rubber is another rubber that is associated with the manufacture of tyres; it is used either as an impermeable liner in tubeless tyres or in the manufacture of inner tubes. Isayev and Feng have continuously devulcanised a resin-cured, unfilled butyl rubber [78] using a 30 mm extruder that had been fitted with two ultrasonic horns. A number of tests were performed on the resulting DR, e.g., gel fraction, crosslink density and Mw. Severe ultrasonic treatment was required to devulcanise the material, but this also led to degradation, with a reduction in Mw and a broadening of the Mw distribution. Good mechanical properties from the revulcanised rubber were obtained, however, the values being comparable with virgin material.

Obtaining a fundamental understanding of how the various rubber types that are present within waste tyres respond to ultrasound devulcanisation, as well as how the presence of fillers such as carbon black influence the process, assisted the Akron team in the application of the technique to waste tyre rubber itself. Work on waste tyre rubber had been going on for a number of years; for example, a paper on the subject was published in 1996 [79], but further work was reported in a more recent paper at an ACS meeting in 2012 [80], which described a study that had been carried out on the devulcanisation of tyre rubber using a new ultrasonic twin-screw extruder. In this study, the effect of particle size was examined by using tyre rubber of two particle sizes (10 and 30 mesh), together with the effects of varying a number of process parameters. To assist in monitoring the process, the die pressure and ultrasonic power consumption were recorded, and the degree of devulcanisation of the samples was investigated by measuring crosslink density, gel fraction and revulcanisation behaviour. The results that were obtained showed, for example, that revulcanisates with a greater degree of devulcanisation exhibited higher elongation at break, while those with a lower degree of devulcanisation exhibited a higher strength and modulus. Also, revulcanisates of the 30-mesh tyre rubber exhibited consistently higher elongation at break values than those from the 10 mesh.

Isayev and the group have also applied the technique to other rubber types. In a recent paper presented at the *ACS Spring 2011 Meeting* [81] they reviewed the development of ultrasound for the devulcanisation of a number of different types of filled and unfilled rubber, including EPDM and silicone, NR and SBR. Its application to the removal of the crosslinks that are present in certain crosslinked plastics, such as crosslinked HDPE, was also included. Earlier papers have been published that have also reported the results obtained on EPDM [82] and silicone rubber [83]. A paper that described the results obtain with polyurethane rubber was published in 2004 [84].

The analysis technique of NMR has been used to characterise the devulcanisation products of ultrasound. Such a study has been carried out and published in a paper in 2007 by Massey and co-workers [85]. In that paper the workers reported that they had used a range of ^{13}C- and ^{1}H-NMR spectroscopic techniques at 70 °C to study the molecular and segmental mobilities in unfilled NR before and after sulfur crosslinking and after subsequent devulcanisation using intense ultrasound under different conditions. The objective of the work was to confirm the molecular mechanisms involved in the devulcanisation of NR using ultrasound. The NMR techniques could not clearly distinguish between entangled and crosslinked network mobility, but unentangled sol (i.e., devulcanised material) and oligomeric species were separable. Ultrasound reactor settings were found to affect the amount of extractable sol generated, with around two-thirds of this being entangled with a Mn above 10,000 Da. The samples were also found to contain about 2 wt% of a low Mw fraction having a Mn of <400 Da, which was not thought to have been generated by the ultrasound process as this is usually ineffective in producing additional oligomeric material. The intramolecular (i.e., backbone) mobility was found to decrease slightly, suggesting that the ultrasound induces isomerisation and cyclisation. The workers also found that with NR, in contrast to other rubbers (e.g., SBR), neither the glass transition nor the sol diffusion rate were greatly affected by exposure to ultrasound. NMR has also been used to characterise other rubbers that have been subjected to ultrasound devulcanisation, including BR [86].

The work at Akron was the subject of an article in *Tire Technology International* in 2012 [87]. In this, reference was made to the fact that both laboratory and semi-industrial scale processes has been set up to devulcanise waste rubber materials such as tyres and industrial off-cuts and scrap rubber into high-quality recyclate that could be blended with virgin rubber to make new products.

4.7 Microwave devulcanisation

4.7.1 Background to process

This process can be regarded as a form of thermal devulcanisation, and the use of microwave energy to devulcanise rubber by causing molecular motion within it, resulting in heat generation, has been assessed by a number of workers. With regard to how the process is carried out in practice, the review published by Myhre and MacKillop [1] stated that a number of the processes that have been developed refer to a Goodyear Tire and Rubber Company patent (US 4,104,205) that was published in 1978.

If the process can be controlled accurately enough, the microwave energy can selectively break the sulfur–sulfur bonds within the crosslinks in sulfur-cured rubbers on the basis of the relative bond energies described in Section 4.2.1.

One requirement of the technique is that rubber compounds must be polar enough to absorb the microwave energy at a high enough rate for it to function effectively. Although this criterion would seem to favour the more polar rubbers (e.g. nitrile rubbers), in practice the technique has been applied to a range of rubber types (Section 4.7.2), and the additives that are present in a rubber compound can often play an important role in influencing the outcome.

One potential drawback of the process is that it can be difficult to control due to the speed at which the temperature of the rubber can increase during the devulcanisation reaction. A cooling unit is often installed at the end of the process to remove heat to reduce the possibility of degradation and assist in post-production handling.

Because of the possibility of over-heating, microwave devulcanisation can sometimes cause problems with diene rubber-based compounds, due to their limited thermal stability. So, although Section 4.7.2 shows that successful work with these materials has been carried out, workers have often tended to concentrate on rubbers that have greater heat resistance, such as butyl rubber and EPDM rubber. The work reported below also shows that these studies have often been successful, with, in the case of butyl rubber, the DR showing a tendency to revulcanise without the need for recompounding.

The use of microwaves in conjunction with a chemical devulcanisation agent is a variation of this technology that assists the process to be more targeted towards the removal of crosslinks. In this variation of the technology, the microwaves are used to generate the heat to enable the devulcanisation reaction to proceed.

4.7.2 Examples of processes and published studies

As discussed in Section 4.2, both monosulfidic and disulfidic bonds are known to be present in the crosslinks in sulfur vulcanised rubbers. To assist in the understanding of how these types of crosslink react to the application of microwave and conductive energy, two model compounds have been synthesised: digeranyl monosulfide and digeranyl disulfide [88]. In addition to these two forms of heating, the influence of the devulcanising agent, diphenyl disulfide, on the devulcanisation was also evaluated in this work. Synthesising these two compounds made it possible to study their reactivity to different sets of conditions independently. One of the main findings was that the DPDS was found to be more effective as a devulcanisation agent when used with microwave heating.

Vega and co-workers [89, 90] have used squalene as a model compound for studies to assess the ability of microwave heating rather than conventional heating to control the devulcanisation of rubber using the devulcanisation agent DPDS. They successfully demonstrated, by GPC and high-performance liquid chromatography analysis of the products of the experiments, that the use of microwave heating was effective in being able to control the reaction temperature of the process, and hence the activity of the diphenyl disulfide, which can vulcanise, as opposed to devulcanise, a diene system under certain conditions. Structural analysis of the products that resulted from the devulcanisation reaction was undertaken and the results showed that the DPDS ended up in a pendant group on the squalene chain. The group also assessed the commercial potential of the technique and concluded that using microwaves was an efficient way of providing heat to enable DPDS to act as a devulcanisation agent, and that it could be used commercially.

An article by Landini and co-workers [91] describes how small samples (250 g) of butyl rubber have been devulcanised by the use of a high-frequency (2.45 GHz) microwave. In this preliminary batch assessment, various powers (e.g., 1,000 and 3,000 W) and heating times (e.g., 9–25 min) were used, and the test results obtained indicated that devulcanisation had occurred and that upon subsequent heating some samples were found to revulcanise without the addition of any crosslinking agents.

A group in China [92] have investigated the non-thermal efficiency of microwaves to devulcanise NR tyre tread. They found that even exposing the tread to microwaves for a short period of time resulted in a sample that had a significantly greater equilibrium swelling index than that of an unexposed sample. Also, the sol fraction in the treated vulcanisate remained reasonably constant with time of exposure, indicating that main-chain breakdown was minimal. The breakdown of the crosslink structure at lower temperatures was thought to result mainly from the non-heating effects and showed that greater levels of crosslink reduction could be achieved with the use of microwaves than with conventional heating. The workers concluded that higher productivity, lower devulcanisation temperatures, less environmental pollution and DR with better physical properties can result from using a microwave process. Another group at Yangzhou University [93] devulcanised waste NR powder using microwaves and blended the devulcanised material with virgin NR to produce vulcanisates. A range of properties for these vulcanisates were then established (e.g., mechanical properties, compression set, swelling ratio and crosslink density) and compared to blends where the waste NR had not been devulcanised.

Scuracchio and co-workers [94] have investigated the effects of various parameters (e.g., exposure time and number of treatment steps) on the microwave devulcanisation of ground waste from tyres. The thermal and rheological properties of the devulcanised samples were evaluated using the techniques of differential scanning calorimetry (DSC), DMA and capillary rheometry, and a measure of the degree of devulcanisation established by Soxhlet solvent extraction. The results obtained showed that significant changes occurred in the DR when it was exposed to microwaves for long periods of time.

In another study, involving ground rubber from passenger tyre treads and sidewalls, Scuracchio and co-workers [95] investigated the thermal properties and thermo-oxidative degradation of the irradiated rubber by DSC and TGA. These tests showed changes in the glass transition temperature and thermo-oxidation behaviour of the rubber after irradiation, and these changes were related by the group to the structure of the DR.

At the Qingdao Institute of Chemical Technology in China [96], workers have studied the influence that a range of rubber sample variables have on the microwave devulcanisation process, and how changes to these affected the physical properties and solvent swelling behaviour of the resulting devulcanisates. The sample variables included different rubber types (NR and SBR), the particle size of the rubber crumb and the type of reinforcing filler. These variables were related to devulcanisation temperature and cycle time. The results showed that a faster increase in temperature was obtained with larger crumb particle sizes, that NR crumb increased in temperature faster than SBR crumb, and that the presence of carbon black caused the temperature to rise quicker than if a silica filler was present. The solvent swelling data showed that the devulcanisation of NR had occurred more effectively than SBR, that rubber crumb (1–3 mm) devulcanised more readily than rubber powder (40–60 mesh), and that the presence of a silica filler encouraged devulcanisation more than a carbon black filler. At the same institute, Zhang and co-workers [97] conducted an investigation into the mechanism of microwave devulcanisation and the revulcanisation process. Their results revealed that revulcanisation was divided into two phases. The first phase consisted of the self-crosslinking of macromolecular polysulfidic radicals and macromolecular cyclic polysulfides that were present in the DR, and the second the re-crosslinking of the rubber chains by reacting with the added crosslinking agent. They also found that the revulcanisation process was very 'scorchy', with little or no induction period present in the curing reaction.

Workers in Brazil [98] have recovered scraps of industrial SBR waste and then, after preparing crumb from them by an ambient grinding process, employed microwaves to devulcanise the rubber. Once devulcanised, the vulcanisation behaviour of the rubber was determined by oscillatory disk rheometry and samples vulcanised with and without a post-cure. The samples were then tested so that their mechanical and crosslink densities could be compared.

A lot of work has also been carried out in Brazil when it comes to using this technique for the devulcanisation of EPDM [99]. Waste EPDM from the automotive sector was exposed to microwave radiation for between 2 and 5 min and the DR produced characterised by a DSC and TGA. The degree of devulcanisation was assessed using gel content measurements. Workers from the same Brazilian university [100] have also devulcanised EPDM rubber by microwaves and then blended it with low-density polyethylene (LDPE) in the presence of a peroxide to improve the interfacial interaction between the two phases. The presence of the devulcanised EPDM in the LDPE matrix resulted in a reduction in the deformation and traction strength, but a significant increase in the elastic modulus values and impact strength. DSC data obtained on the mixture showed that the presence of the EPDM had not altered the crystallisation and fusion enthalpies of the LDPE. The devulcanisation process was found to significantly reduce the gel content of the EPDM, demonstrating that it had been successful in removing a large number of crosslinks. In a continuation of the studies on EPDM at this university, a recent paper by Pistor and Zattera [101] examined the efficiency of devulcanising EPDM using microwaves by varying the exposure from 2 to 4 min. SEM and TGA using different heating rates were then used to determine the kinetic degradation parameters according to the Flynn–Wall–Ozawa and Criado methods. The results showed that exposure to the microwave radiation for up to 4 min could be used without exceeding the degradation temperature of the EPDM. The group has also carried out work to evaluate the influence that the presence of paraffinic oil in the EPDM has on the process [102]. Industrial EPDM automotive waste was extracted to remove the oil and the vulcanisate was then exposed to microwaves for periods ranging between 2 and 5 min. Characterisation work on the devulcanised samples, both extracted and un-extracted, involved DSC, TGA and the determination of gel content. The gel content results indicated that the paraffin oil affects the devulcanisation process, while the DSC revealed that the glass transition temperature of the EPDM decreased with increasing exposure to the microwaves and the presence of a thermal transition that was characteristic of uncured rubber was observed. The TGA results showed that, for certain levels of conversion, the activation energy of the degradation process decreased with an increase in microwave exposure time.

At Pisa University, work has also been carried out on EPDM by Bani and coworkers [103]. Carbon-black-filled EPDM crumb samples were devulcanised by exposing them to microwaves so that their temperature increased to over 300 °C. This process, in a short period of time, produced a rubber that had a relatively low crosslink density and could produce high-quality blends with virgin EPDM due to good interfacial adhesion. Once vulcanised, these blends were found to have superior physical properties to those produced using untreated EPDM crumb.

4.8 Microbiological devulcanisation

4.8.1 Background to process

In this process, which has been in existence as a concept for at least 30 years and is often referred to as a microbial process, chemolithiotrophic bacteria in aqueous suspensions in the presence of oxygen can be used to selectively attack the sulfur crosslinks in the surface of waste rubber crumb particles. The process can also be regarded as a type of biodegradation process, the by-products of which can include elemental sulfur, sulfates, sulfides and sulfuric acid. Specific examples of the bacteria that have been used to devulcanise rubber in this way include species of *Thiobacillus*, e.g., *T. thiooxidans*.

One estimate of the depth to which these devulcanisation reactions penetrate into the surface of the rubber particles is <1 μm [104]. The process can also be regarded, therefore, as a technique for surface modification (other types of surface modification and activation are reviewed in Chapter 6, Section 6.6), which, by desulfurisation, activates the surface of the rubber particles and improves their interfacial bonding within the matrix of the rubber compounds into which they are incorporated, hence enhancing the quality of the subsequent products.

Because it is primarily a surface phenomenon, bacterial devulcanisation often requires finely ground rubber powder (e.g., 200 mesh) to be most effective, as a high surface area to volume ratio is needed, and the reaction is usually carried out in a temperature-controlled bioreactor. Potential disadvantages of the process, therefore, can include the relatively high cost of the small-particle-size rubber crumb, the relatively high capital outlay, and the fact that the rubber crumb has to go through a post-treatment wash and drying process.

4.8.2 Examples of processes and published studies

An example of this type of process has been developed by the Scottish company RecyclaTech. Their process was the subject of a patent in 2003 and used the results of research that had been carried out at Napier University in Edinburgh. One claim of this company is that their process can raise the amount of recycled rubber that can be used in a tyre from 5 to 25%. The process can remove over 20% of the 1.5–3% of sulfur that is present in waste rubber, which renders the rubber suitable for mixing with virgin rubber in relatively large quantities to produce high-value products. Wear tests for Toyota have been conducted that compared a new tread compound containing 10% partially DR with one made using 100% virgin rubber, and there was no discernable difference in wear, suggesting that partially DR produced using the microbial devulcanisation route will perform as well as virgin rubber when used in automotive tyres. This company collaborated with the Waste Resources Action Programme (WRAP) to produce one of their case studies in 2007 [105] which provided an overview of microbial devulcanisation and summarised the technical, economic and environmental benefits (Figure 4.8). Also, an article in the *European Rubber Journal* [106] gave a brief description of the process and mentioned that it was possible to re-use waste tyre rubber that had been devulcanised using this technology in the production of new tyres and in other products, such as basic underlay for carpets, soundproofing materials and wellington boots. The process has also been covered in other trade journals; for example, a small report was published in *Scrap Tire News* [107].

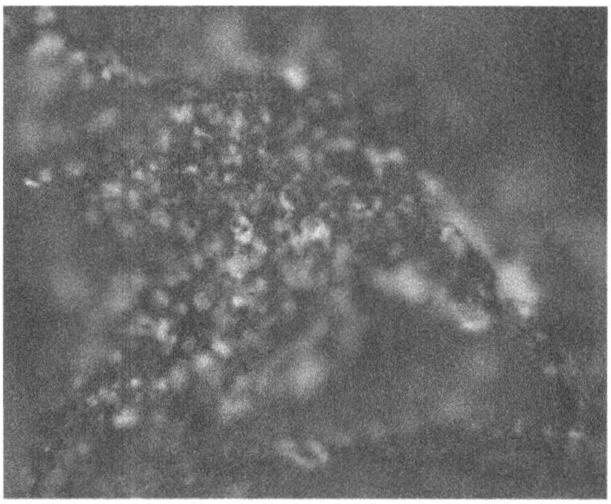

Figure 4.8: Fine rubber particles suspended in water and aerobically treated with desulfurising bacteria – WRAP DART project. Reproduced with permission from WRAP, Banbury, UK. ©WRAP. **Note:** WRAP does not warrant that the functions contained in the material contained in this Site will be uninterrupted or error free, that defects will be corrected, or that this Site or the server that makes it available are free of viruses or bugs or represents the full functionality, accuracy, and reliability of the materials.

Other work that has been carried out with microbial devulcanisation includes use of the microorganism *Pyrococcus furiosus* to carry out devulcanisation under anaerobic conditions. This type of bacteria reduces the sulfur in the crosslinks to hydrogen sulfide [108]. Another type of bacteria, *Sulfolobus acidocaldarius*, can be used to selectively cleave carbon–sulfur bonds to generate sulfate compounds that stay in the rubber and so the sulfur remains available for any revulcanisation treatments [109].

Recent Chinese papers [110, 111] have described how the growth of *Thiobacillus thioparus* with varying amounts of vulcanised NR has been investigated. The influence that the bacteria had on the devulcanisation of the material, and hence its chemical structure and crosslink density, was determined using a range of analytical techniques, including FTIR spectroscopy, X-ray photoelectron spectroscopy (XPS), NMR spectroscopy, and by the analysis of sulfate ions (SO_4^{2-}) in a culture medium. These workers also investigated the effect that incorporating this devulcanised NR crumb as a filler into a virgin NR compound had on its physical properties, such as Shore hardness, tensile strength and modulus at 300%. These blends were also characterised by SEM.

A lot of work in this field has been carried out by workers at the Beijing University of Chemical Technology, and some of their studies are covered below. They carried out the microbial desulfurisation of GTR and ground SBR rubber by *Sphingomonas* sp., which had been obtained from coal mine soil and had the ability to oxidise sulfur [112, 113]. In the case of the GTR [112], it was immersed in the medium for 20 days and the growth curve of the *Sphingomonas* sp. was determined along with the chemical groups that remained on the rubber surface. This analysis showed a reduction in both the level of sulfur and carbon–carbon double bonds. Both desulfurised rubber particles and untreated particles were compounded into SBR compounds, and a comparative determination of the vulcanisates was undertaken, the results of which showed that samples with the desulfurised rubber in them had better physical properties and higher solvent swelling values. In the case of the ground SBR rubber, an analysis of the desulfurised rubber was carried out using the chemical analysis techniques attenuated total internal reflectance FTIR and XPS; the rubber was incorporated into NR compounds and the crosslink densities, mechanical properties and fracture surface morphologies were determined.

The team at Beijing University of Chemical Technology also studied the effect on NR of using *Thiobacillus ferrooxidans* that had been obtained from the soil associated with an iron ore mine [114]. The rubber crumb was exposed to the microbes for 30 days and then compounded into NR compounds at levels that varied from 10 to 40 phr. The results obtained were similar to the *Sphingomonas* sp. work in that compounds containing the desulfurised crumb had better physical properties than control samples, and chemical group analysis of the crumb surface showed loss of sulfur, thought to be due to oxidation, and loss of carbon–carbon double bonds.

In an earlier study, the group at Beijing used a mercapto-containing yeast to devulcanise ground NR [115], and then characterised a number of the properties of the devulcanised NR, including its mechanical properties, swelling behaviour in solvents, crosslink density, and the amount of sulfur present on the surface of the particles.

The different aerobic microbiological processes that have been developed for the desulfurisation of ground waste rubber where the sulfur in the crosslinks is oxidised to sulfate compounds have been described in a small review published by

Kadam and Mhaska [116]. Earlier reviews on the subject have been published by Christiansson and co-workers [117] and Holst and co-workers [118].

4.9 Miscellaneous devulcanisation processes

This section contains processes that fit into the following categories:

- Processes that, from the information available, do not easily fit into any of the sections above or are sufficiently different to warrant separate treatment.
- Processes for which sufficient information may not be available to place them in one of the sections above because they are proprietary.

The processes that fall into these categories are covered below. They have not been placed in any particular order.

SRI Elastomers Sdn Bhd in Malaysia has developed a technique that activates 40-mesh rubber crumb so that when it is blended into virgin rubber compounds it can be used to produce high-quality products [119]. One such product that has been produced in this way from internal factory waste is a medium-grade retread compound. The SRI activated 40-mesh crumb was blended into virgin compound at levels up to 15% w/w with no significant change in tensile strength, abrasion loss, rebound resilience and tear resistance, and less than 10% loss in elongation at break. When the SRI activation process was used on waste tyre rubber crumb to produce a number of medium-grade truck retread compounds, good quality blends could be produced with levels of the SRI compound up to 20%. Only relatively small changes (up to ~15% reduction) in tensile strength and elongation at break were found at this level of replacement, and the abrasion loss decreased by up to ~20%. Another example that was shown in this presentation was that of a commercial motorcycle tread compound. At a replacement level of 20%, the SRI compound was found to reduce the elongation at break values up to ~22% and the tensile strength up to 20%, but there was an improvement in abrasion loss of up to 15%. SRI is marketing its product globally to both new tyre manufactures and tyre retreading companies. It was reported in January 2011 [120] that SRI was producing up to 27,000 tonnes per year of DR and the product was being used to produce light commercial vehicle tyres in cooperation with a Malaysian tyre manufacturer.

Another process that is also used on waste rubber crumb to enable it to be incorporated at relatively high levels into new compounds to produce high-quality rubber products is a patented process developed by Polymer Recyclers Ltd (PRL). The PRL process involves high shear mixing and, although according to Kind of PRL [121] no direct attempt is made to devulcanise the rubber, some may take place during the process. A schematic diagram showing the PRL manufacturing process is shown in Figure 4.9, and a photograph of ingredients being loaded into the high-shear internal mixer is provided in Figure 4.10. Some of the results obtained using

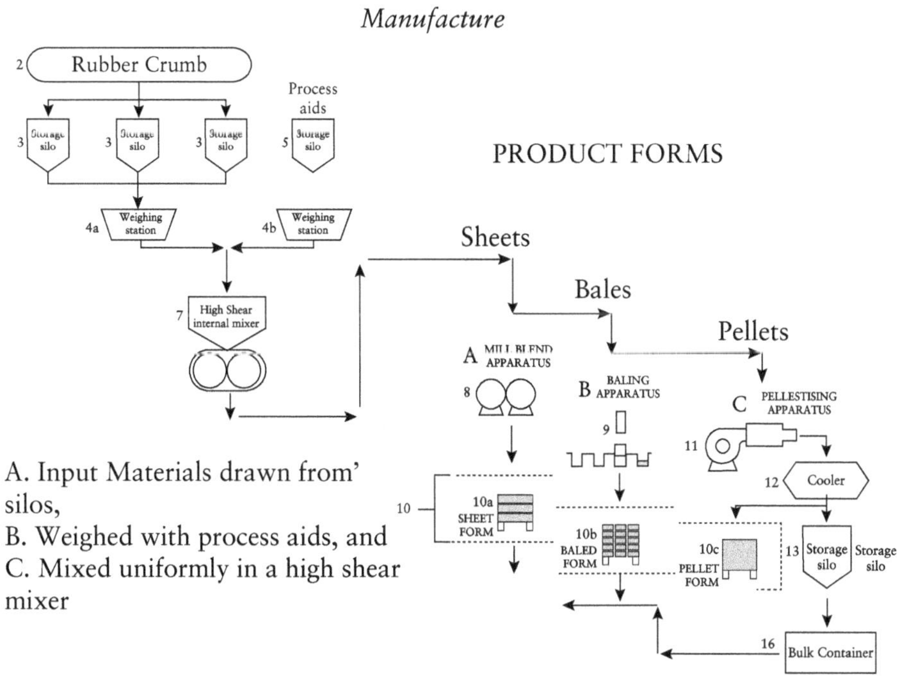

Figure 4.9: Schematic diagram of the PRL manufacturing process for the production of new products from waste rubber crumb. Reproduced with permission from Polymer Recyclers Ltd, Dordon, Tamworth, UK. ©2014, Polymer Recyclers Ltd.

the PRL product in new rubber compounds, together with a comparison of these data with those obtained using two devulcanisation technologies (Levgum and De-Link), are shown in Chapter 5, Section 5.2.2.6.

A Russian group [122] have carried out a study that involved reclaiming butyl rubber from waste tyre curing bags using cobalt-60 gamma radiation at a dose of 50 kGy. The reclaimed butyl rubber was then revulcanised using three different cure systems: *p*-benzoquinone dioxime/manganese dioxide, *p*-dinitrosobenzene and quinol ether. The dinitroso-generating system was the most effective at revulcanisation, and the compounds that resulted had a greater level of resistance to heat ageing than analogous butyl rubber vulcanisates. The workers considered this effect to be due to the inhibition of the thermo-oxidative degradation process by breakdown products of alkylphenyl formaldehyde resin (the original curative), which were formed by irradiation of the rubber.

A group of workers drawn from the Universities of Kocaeli and Hacettepe [123] have carried out a comparative study of the results that were obtained from waste butyl rubber crumb that had been devulcanised using conventional methods and end-of-life butyl rubber inner tubes that had been irradiated using a dose of 120

Figure 4.10: Loading ingredients into the high-shear internal mixer in the PRL process. Reproduced with permission from Polymer Recyclers Ltd, Dordon, Tamworth, UK. ©2014, Polymer Recyclers Ltd.

kGy of gamma rays. The study involved incorporating 15 phr of each source of recycled butyl rubber back into a new butyl rubber compound and comparing the mechanical properties, rheological properties and the degree of carbon black dispersion with those provided by a pure virgin butyl rubber compound. The workers concluded that, although the use of both forms of recycled butyl caused a reduction in properties, the degree of reduction that was observed with the irradiated butyl rubber was less marked. The irradiated butyl rubber was therefore shown to be compatible with the virgin butyl rubber compound, so the workers considered that this recycling route was a technically viable option for the re-use of waste butyl rubber products.

Fukumori and Sato [124, 125] reported on the development of continuous reactive processing methods for recycling waste rubber that have been developed at the Toyota Central Research and Development Laboratories in Japan. The technique, which uses a twin-screw extruder with pulverising and devulcanising sections, and also has a novel deodorising capability, can recycle rubber waste generated from both the automotive and GRG sectors. The devulcanisation of the waste rubber occurs due to a combination of shear stress, temperature and pressure and utilises, as a number of processes do, the differences in both the elastic constants and the bond energies between the carbon–carbon main-chain bonds and the sulfur–sulfur crosslink bonds (Section 4.2). When used in the devulcanisation mode, the technique has been shown to be capable of devulcanising automotive EPDM weather strip waste, which, when compounded with a conventional accelerated sulfur cure system, was re-cured into a new product that had physical properties that were almost comparable to the virgin material. This technology has been used to recycle EPDM waste generated during automotive manufacturing operations in a plant with a production capacity of 500 tonnes per year [126]. The recycled rubber produced by the plant has been used to produce various automobile rubber products, e.g., head-lamp seals, radiator support seals, boot weather strip, and hose protectors. This system has also been used to devulcanise tyre rubber, which was then incorporated, at a level of 10%, into a tread formulation. A comparative tread wear test over 200,000 km of this product against a 100% virgin product showed that tread wear was comparable.

This Toyota continuous devulcanisation process can also be incorporated into a more complex continuous single process that enables both dynamic devulcanisation and dynamic vulcanisation to be carried out. The twin-screw extruder in this system has specific areas for the following operations:
- EPDM waste feed zone
- Pulverisation
- Devulcanisation with deoderisation
- Blending zone with feeds for PP resin pellets (start of zone) and additives (end of zone)
- Dynamic crosslinking zone for EPDM fraction
- Die of extruder for shaping extrudate and then cooling bath and pelletiser

Devulcanising rubber can result in products that have bad odours, and this has been addressed in this system by the use of a novel deodorisation process that involves the injection of high-pressure water into the extruder. Odour-producing compounds are trapped in the high-pressure water vapour and removed *via* vents.

The combined system of processes that is described above is capable, once the design has been optimised to achieve the desired morphology, of turning waste

EPDM (80 wt%) and PP (20 wt%) into a TPE in which the crosslinked EPDM (major component and in domains of 1 μm or less in size) is dispersed within a PP matrix (minor component). The TPE pellets obtained by the system are capable of being extrusion-moulded or injection-moulded like conventional thermoplastics, and the products produced by these processes have tensile properties, processing properties and a surface appearance that is comparable to commercial TPE. Mass production of this TPE has been carried out and has been evaluated in the production of automotive products, such as vacuum sensing hoses, sunroof drain hoses, door glass runs and outer belt line seals.

The use of ultrasound by the group at Akron University to devulcanise rubber has been covered in Section 4.6. The same group of workers have also used the technique, with a specially designed reactor, to treat PP/GTR blends to produce novel materials [127]. In the reactor, two horns were placed in a slit die attached to a plastic extruder. PP/GTR blends missed using a twin-screw extruder were passed through the reactor to devulcanise the GTR, and the blends then dynamically revulcanised using an internal mixer. The results of the subsequent testing work showed that this process created blends that had better mechanical properties than those generated by previous experiments conducted by the team.

A company called Revultec have developed a devulcanisation system called the Revultec system [128], which uses a relatively inexpensive solvent, such as hexane, in its supercritical state to devulcanise waste tyre crumb rubber. At the end of the process, up to 98% of the solvent can be recovered for re-use, and it is claimed that the DR is of a sufficiently high quality to make new rubber products, including tyres.

References

1. M. Myhre and D.A. MacKillop, *Rubber Chemistry and Technology*, 2002, **75**, 2002, 429.
2. M. Myhre, S. Saiwari, W. Diekes and J. Noordermeer, *Rubber Chemistry and Technology*, 2012, **85**, 3, 408.
3. *Evaluation of Waste Tire Devulcanisation Technologies*, Report by CalRecovery Inc. for the Integrated Waste Management Board of California, December 2004.
4. S. Majumdar, *Chemical Weekly*, 2009, **54**, 48, 200.
5. S. Majumdar, *The Rubber International Magazine*, 2008, **10**, 11, 25.
6. W.J. MacKnight and A.V. Tobolsky in *Polymeric Sulfur and Related Polymers*, Interscience Publishers, New York, NY, USA, 1965.
7. K. Murakami and K. Ono in *Chemorheology of Polymers*, Elsevier, New York, NY, USA, 1979.
8. E. Finazzi, A. Gallo and P. Lucci, *Rubber World*, 2011, **243**, 21, 21.
9. D.T. Chen, C.A. Perman, M.E. Riechert and J. Hoven, *Journal of Hazardous Materials*, 1995, **44**, 53.
10. M. Kojima, M. Tosaka, E. Funami, K. Nitta and M. Ohshima, *The Journal of Supercritical Fluids*, 2005, **35**, 3, 175.

11. M. Kojima, K. Ogawa, H. Mizoshima, M. Tosaka and S. Kohjiya, *Rubber Chemistry and Technology*, 2003, **76**, 4, 957.
12. M. Kojima, M. Tosaka, Y. Ikeda and S. Kohiya, *Journal of Applied Polymer Science*, 2005, **95**, 1, 137.
13. M.A.L. Verbruggen, L. Van Der Does, J.W.M. Noordermeer, M. Van Duin and H.J. Manual, *Rubber Chemistry and Technology*, 1999, **72**, 4, 731.
14. X.X. Zhang, C.H. Lu and M. Liang, *Plastics, Rubber and Composites*, 2007, **36**, 7–8, 370.
15. B. Maridass and B. Gupta, *Journal of Elastomers and Plastics*, 2006, **38**, 211.
16. M. Balasubramanian, *Journal of Polymer Research*, 2009, **16**, 2, 133.
17. W. Rui, M. Guobing, W. Dun, L. Huayong, X. Zhengjun and T. Guoliang, *China Synthetic Rubber Industry*, 2010, **33**, 4, 313.
18. Z. Kaijun, W. Rui, L. Chunlin, L. Huayong and T. Guoliang, *China Synthetic Rubber Industry*, 2010, **33**, 3, 171.
19. X. Chen, Z. Xu and Y. Zhang, *Polymer Materials Science and Engineering*, 2010, **26**, 2, 91.
20. A. Macsiniuc, A. Rochette and D. Rodrigue, *Progress in Rubber, Plastics and Recycling*, 2010, **26**, 2, 51.
21. Z. Liang, L. Qiuying and Wu Chifei, *China Synthetic Rubber Industry*, 2010, **33**, 2, 142.
22. D. Brown in *Proceedings of the Mining and Metals Recycling of Rubber Meeting*, Institute of Materials, London, UK, 2008.
23. W. Maddever and W. Brown in *Proceedings of the ACS Rubber Division Meeting – Fall 2010*, Milwaukee, USA, American Chemical Society, Washington, DC, USA, 2010, Paper No.46.
24. D. Brown, M. Burgoyne and W. Maddever in *Proceedings of the ACS Rubber Division Meeting – Fall 2009*, Pittsburgh, USA, American Chemical Society, Washington, DC, USA, 2009, Paper No.159.
25. D. Shaw, *European Rubber Journal*, 2010, **192**, 3, 16.
26. D. Shaw, *European Rubber Journal*, 2009, **191**, 1, 30.
27. M. Meysami and C. Tzoganakis in *Proceeding of the 67*th *SPE Annual Technical Conference*, Chicago, USA, Society of Plastics Engineers, Bethel, CT, USA, 2009, p.640.
28. D. Shaw, *European Rubber Journal*, 2010, May/June, p.21.
29. J. Shen and Y. Zhang, *Polymer Materials Science and Engineering*, 2009, **25**, 1, 118.
30. B. Maridass and B. Gupta, *Polymer Composites*, 2008, **29**, 12, 1350.
31. D. Kang, J. Mead, Y. Choi and R. Farris, *Rubber World*, 2007, **237**, 3, 24.
32. H. Yazdani, M. Karribi, I. Ghasmi, H. Azizi and G.R. Bakhshandeh, *Journal of Vinyl and Additive Technology*, 2011, **17**, 1, 64.
33. C. Tzoganakis and Q. Zhang in *Proceedings of the Global Plastics Environmental Conference*, Detroit, MI, USA, Society of Plastics Engineers Environmental Division, Lindale, GA, USA, 2004, Paper No.49.
34. D. Shaw, *European Rubber Journal*, 2010, **192**, 3, 21.
35. J. Savage and Q. Hartley in *Proceedings of High Performance Elastomers*, Frankfurt, Germany, 5–6th December, Smithers Rapra, Shawbury, UK, 2007, Paper No.12.
36. M. Myhre, S. Saiwari, W. Dierkes and J. Noordermeer, *Rubber Chemistry and Technology*, 2012, **85**, 3, 408.
37. B. Saville and A.A. Watson, *Rubber Chemistry and Technology*, 1967, **40**, 100.
38. R.J. Crenlyn in *An Introduction to Organosulfur Chemistry*, Wiley, New York, NY, USA, 1996.
39. *Chemistry of Organosulfur Compounds*, Ed., L.I. Belen'kii, Ellis Horwood Ltd, Chichester, UK, 1990.
40. PTA News Bureau, *Polymer and Tyre Asia*, 2011, April/May, 66.
41. *European Rubber Journal*, 2009, **191**, 2, 6.

42. S. Nieto in *Proceedings of the ACS Rubber Division Meeting – Fall 2009*, Pittsburgh, USA, American Chemical Society, Washington, DC, USA, 2009, Paper No.160.
43. S. Goodchild, *Tyres and Accessories*, 2008, December, p.74.
44. Anon., *Proceedings of the ACS Rubber Division Meeting – Fall 2009*, Pittsburgh, USA, American Chemical Society, Washington, DC, USA, 2009, Paper No.156.
45. M.J. Forrest in *Proceedings of the RubberChem Conference*, Vienna, Austria, Smithers Rapra Technology Ltd, Shawbury, UK, 2010, Paper No.17.
46. M.E. Antonio, *Tire Technology International*, 2002, p.86.
47. G.K. Jana and C.K. Das, *Progress in Rubber, Plastics and Recycling Technology*, 2005, **21**, 4.
48. G.K. Jana and C.K. Das, *Polymer Plastics Technology and Engineering*, 2005, **44**, 1399.
49. S.J. Asadauskas, E. Kaminskas, T. Langaitis and A. Jukna in *Proceedings of the Tire Technology Expo 2014*, 11–13th February, Cologne, Germany, 2014.
50. K.A.J. Dijkhuis, I. Babu, J.S. Lopulissa and W.K. Dierkes in *Proceedings of the ACS Rubber Division – Fall 2006*, Cincinnati, USA, American Chemical Society, Washington, DC, USA, 2006, Paper No.24.
51. P. Sutanto, F.L. Laksmana, F. Picchioni and L.P.B.M. Janssen, *Chemical Engineering Science*, 2006, **61**, 6442.
52. Debapriya De, Debasish De and G.M. Singharoy, *Polymer Engineering and Science*, 2007, **47**, 7, 1091.
53. K.A. Dukhuis, P. Sutanto, W.K. Dierkes and J.W.M. Noordermeer, *Gummi Fasern Kunststoffe*, 2010, **63**, 2, 92.
54. K.A. Dukhuis, I. Babu, J.S. Lopulissa, J.W.M. Noordermeer and W.K. Dierkes, *Gummi Fasern Kunststoffe*, 2009, **62**, 12, 715.
55. K.A. J. Dijkhuis, J.W.M. Noordermeer and W.K. Dierkes in *Proceedings of the ACS Rubber Meeting – Fall 2008*, Louisville, USA, American Chemical Society, Washington, DC, USA, 2008, Paper No.14.
56. M.A.L. Verbruggen, L. Van Der Does, J.W.M. Noordermeer and M. Van Duin, *Journal of Applied Polymer Science*, 2008, **109**, 2, 976.
57. N. Roy, S. Kumar and C.K. Das, *Elastomery*, 2009, **13**, 6, 20.
58. G. Youyong, L. Xiaolin, X. Xiao and W. Shemao, *China Synthetic Rubber Industry*, 2010, **33**, 5, 370.
59. A.R. Jalilvand, I. Ghasemi and H. Azizi, *Progress in Rubber, Plastics and Recycling Technology*, 2008, **24**, 1, 33.
60. Z. Xiankui, G. Lei, B. Huiguang and W. Chuansheng, *China Synthetic Rubber Industry*, 2009, **32**, 4, 321.
61. L.V. Popova, O.V. Karmanova, S.G. Tikhomirov and S.I. Korystin, *Kauchuk i Rezina*, 2007, **5**, 28.
62. A.I. Isayev and S. Ghose in *Rubber Recycling*, Eds., S.K. De, A.I. Isayev and K. Khait, CRC Press, Boca Raton, FL, USA, 2005, Chapter 9.
63. A.I. Isayev, S.P. Yushanov, D. Schworm and A. Tukachinsky in *Proceedings of the ACS Rubber Division Meeting – Fall 1995*, Cleveland, USA, American Chemical Society, Washington, DC, USA, 1995, Paper No.123.
64. S.P. Yushanov, A.I. Isayev and V.Y. Levin, *Journal of Polymer Science*, 1996, **34**, 14, 2409.
65. S.P. Yushanov, A.I. Isayev and S.H. Kim in *Proceedings of the 150*th *ACS Rubber Division Meeting – Fall 1996*, Louisville, USA, American Chemical Society, Washington, DC, USA, 1996, Paper No.59.
66. A.I. Isayev, J. Chen and A. Tukachinsky, *Rubber Chemistry and Technology*, 1995, **68**, 2, 267.
67. J. Yun, J.S. Oh and A.I. Isayev in *Proceedings of the 150*th *ACS Rubber Division Meeting – Fall 1996*, Cincinnati, USA, American Chemical Society, Washington, DC, USA, 2000, Paper No.113.

68. A.I. Isayev, S.H. Kim and V. Yu Levin in *Proceedings of the 150*th *ACS Rubber Division Meeting – Fall 1996*, Louisville, USA, American Chemical Society, Washington, DC, USA, 1996, Paper No.62.
69. V. Yu Levin, S.H. Kim and A.I. Isayev in *Proceedings of the 150*th *ACS Rubber Division Meeting – Fall 1996*, Louisville, USA, American Chemical Society, Washington, DC, USA, 1996, Paper No.60.
70. A.I. Isayev, S.H. Kim and W. Feng, *Rubber Chemistry and Technology*, 2005, **78**, 4, 606.
71. V. Yu Levin, S.H. Kim, A.I. Isayev, J. Massey and E. von Meerwall, *Rubber Chemistry and Technology*, 1996,**69**, 1, 104.
72. C.K. Hong and A.I. Isayev, *Journal of Applied Polymer Science*, 2001, **79**, 13, 2340.
73. K.H. Chang and A.I. Isayev, *Journal of Material Science*, 2002, **37**, 2, 385.
74. S. Ximei and A.I Isayev, *Rubber Fibres Plastics International*, 2009, **4**, 3, 130.
75. S. Ximei and A.I Isayev, *Gummi Fasern Kunststoffe*, 2009, **62**, 5, 291.
76. S. Ximei and A.I Isayev, *Rubber Chemistry and Technology*, 2008, **81**, 1, 19.
77. J. Seok Oh and A.I. Isayev, *Journal of Applied Polymer Science*, 2004, **93**, 3, 1166.
78. W. Feng and A.I. Isayev, *Journal Applied Polymer Science*, 2004, **94**, 3, 1316.
79. A. Tukachinsky, D. Schworm and A.I. Isayev, *Rubber Chemistry and Technology*, 1996, **69**, 1, 92.
80. A.I. Isayev, T. Liang and T.M. Lewis in *Proceedings of the ACS Rubber Division Meeting – Fall 2012*, Pittsburgh, USA, American Chemical Society, Washington, DC, USA, 2012, Paper No.35.
81. A.I Isayev in *Proceedings of the ACS Rubber Meeting – Spring 2011*, Akron, OH, USA, American Chemical Society, Washington, DC, USA, 2011, Paper No.9.
82. Y. Jushik and A.I. Isyaev in *Proceedings of the ACS Rubber Division Meeting – Fall 2001*, Cleveland, OH, USA, American Chemical Society, Washington, DC, USA, 2001, Paper No.46.
83. B. Diao, A.I. Isayev and V.Yu. Levin in *Proceedings of the ACS Rubber Division Meeting – Fall 1997*, Cleveland, OH, USA, American Chemical Society, Washington, DC, USA, 1997, Paper No.96.
84. S. Ghose and A.I. Isayev, *Journal of Elastomers and Plastics*, 2004, **36**, 3, 213.
85. J.L. Massey, J.C. Parr, T.A. Wagler, E. von Meerwell, C.K. Hong and A.I Isayev, *Polymer International*, 2007, **56**, 7, 860.
86. J.S. Oh, A.I. Isayev and E. von Meerwall, *Rubber Chemistry and Technology*, 2004, **77**, 4, 745.
87. K. Read, *Tire Technology International*, 2012, October, 42.
88. B. Vega, C. Kraushaar, N. Agullo and S. Borros, *Kautschuk Gummi Kunststoffe*, 2008, **61**, 7–8, 390.
89. B. Vega, L. Montero, S. Lincoln, N. Agullo and S. Borros, *Journal of Applied Polymer Science*, 2008, **108**, 3, 1969.
90. B. Vega, L. Montero, E. Lincoln, N. Agullo and S. Borros in *Proceedings of the ACS Rubber Division Meeting – Fall 2006*, Cincinnati, OH, USA, American Chemical Society, Washington, DC, USA, 2006, Paper No.8.
91. L. Landini, S.G. De Araujo, A.B. Lugao and H. Wiebeck, *European Polymer Journal*, 2007, **43**, 6, 2725.
92. T. Wei-qiang, L. Jun-jie and Z. Hai, *China Rubber Industry*, 2006, **53**, 8, 453.
93. M. Caiyun, L. Ping, M. Limin, H. Hui and Y. Chan, *China Synthetic Rubber Industry*, 2010, **33**, 5, 366.
94. C.H. Scuracchio, D.A. Waki and R.E.S. Bretas, *Polimeros: Ciencia e Tecnologia*, 2006, **16**, 1, 46.
95. C.H. Scuracchio, D.A. Waki and M.L.C.P. Da Silva, *Journal Thermal Analysis and Calorimetry*, 2007, **87**, 3, 893.
96. Z. Shugao, Z. Ping, C. Yonghua, L. Xiaohong and Z. Junxue, *China Rubber Industry*, 1999, **46**, 5, 292.

97. P. Zhang, S-G. Zhao and L. Cui, *China Rubber Industry*, 2001, **48**, 8, 458.
98. A. Zanchet, L.N. Carli, M. Giovanela, J.S. Crespo, C.H. Scuracchio and R.C.R. Nunes, *Journal of Elastomers and Plastics*, 2009, **41**, 6, 497.
99. V. Pistor, F.G. Ornaghi, R. Fiorio, A.J. Zattera, P.J. Oliveira and C.H. Scuracchio, *Polimeros: Ciencia e Tecnologia*, 2010, **20**, 3, 165.
100. V. Pistor, F.G. Ornaghi, R. Fiorio and A.J. Zattera, *Journal of Elastomers and Plastics*, 2010, **42**, 5, 417.
101. V. Pistor and A.J. Zattera, *Journal of Elastomers and Plastics*, 2014, **46**, 1, 69.
102. V. Pistor, C.H. Scuracchio and P.J. Oliveira, *Polymer Engineering and Science*, 2011, **51**, 4, 697.
103. A. Bani, G. Polacco and G. Gallone, *Journal of Applied Polymer Science*, 2011, **120**, 5, 2904.
104. M. Loffler and W. Neumann, *Kautschuk Gummi Kunststoffe*, 1995, **48**, 454.
105. *Waste Tyres Case Study – Summary: TYR0008 – DART Microbial Devulcanisation*, Waste Resources Action Programme, Banbury, UK, 2007.
106. H. Gilbert, *European Rubber Journal*, 2008, **190**, 5, 16.
107. *Scrap Tire News*, 2008, **22**, 10, 14.
108. K. Bredberg, J. Persson, M. Christiannson, B. Stenburg and O. Holst, *Applied Microbiological Biotechnology*, 2001, **55**, 1, 43.
109. R.A. Romine, M.F. Romine and L. Snowden-Swan in *Proceedings of the ACS Rubber Division Meeting – Fall 1995*, Cleveland, OH, USA, American Chemical Society, Washington, DC, USA, 1995, Paper No.56.
110. J. Guangming, Z. Suhe, L. Jingyuan and W. Yaq, *China Synthetic Rubber Industry*, 2010, **33**, 6, 449.
111. J. Guangming, Z. Suhe, L. Jingyuan and W. Yaq, *China Synthetic Rubber Industry*, 2010, **33**, 4, 271.
112. Y. Li, S. Zhao and Y. Wang, *Journal of Polymers and the Environment*, 2012, **20**, 2, 372.
113. G. Jiang, S. Zhao, W. Li, J. Luo, Y. Wang, Q. Zhou and C. Zhang, *Polymers for Advanced Technologies*, 2011, **22**, 12, 2344.
114. Y. Li, S. Zhao and Y. Wang, *Polymer Degradation and Stability*, 2011, **96**, 9, 1662.
115. Q. Liusha, Z. Suhe, W. Yaqin and Y. Xiumei, *China Synthetic Rubber Industry*, 2008, **31**, 1, 36.
116. P.G. Kadam and S.T. Mhaska, *Popular Plastics and Packaging*, 2012, **57**, 5, 29.
117. M. Christiansson, O. Holst and B. Stenburg in *Microbial Desulfurisation of Rubber – A Literature Study*, AFR Report No.119, AFN, Naturvårdsverket, Swedish Environmental Protection Agency, Stockholm, Sweden, May, 1996.
118. O. Holst, B. Stenburg and M. Christiansson, *Biodegradation*, 1998, **9**, 301.
119. G.B Sekhar in *Proceedings of the Tire Technology Expo 2014*, 11–13th February, Cologne, Germany, 2014.
120. *Tyres and Accessories*, 2011, 1, 72.
121. R.C. Kind in *Proceedings of RubberChem Conference*, Vienna, Austria, Smithers Rapra Technology Ltd, Shawbury, UK, 2010, Paper No.18.
122. R.R. Vagizova, Y.N. Khakimullin, T.V. Makarov, S.I. Vol'fson, P.A. Stepanov and F.M. Palyutin, *International Polymer Science and Technology*, 2008, **35**, 3, T/7–9.
123. B. Karaagac, M. Sen, V. Deniz and O. Guven, *Nuclear Instruments and Methods in Physics Research*, 2007, **B265**, 1, 270.
124. K. Fukumori and N. Sato in *Proceedings of the International Rubber Conference*, Yokohama, Japan, International Rubber Conference Organisation, Institute of Materials, Minerals and Mining, London, UK, 2005, Paper No.7.
125. K. Fukumori, M. Matsushita, M. Mouri, H. Okamoto, N. Sato, K. Takeuchi and Y. Suzuki, *Kautschuk Gummi Kunststoffe*, 2006, July/August, 405.

126. S. Otsuka, M. Owaki, Y. Suzuki, H. Honda, K. Nakashima, M. Mouri and N. Sato, *SAE Technical Paper 2001-01-0015*, SAE International, Warrendale, PA, USA, 1998.
127. S. Jeong and A.I. Isayev in *Proceedings of the ACS Rubber Division Meeting – Fall 2001*, Cleveland, OH, USA, American Chemical Society, Washington, DC, USA, 2001, Paper No.11.
128. K.J. Fryer in *Tire Technology International Annual Review*, Tire Technology International, Dorking, UK, 2010, p.94.

5 Characterisation of devulcanised rubber and products containing devulcanised rubber

5.1 Characterisation of devulcanised rubber

The large range of processes and technologies that have been developed for the devulcanisation of waste rubber have been described in Chapter 4. One thing that all these processes have in common is the need to characterise and evaluate the product that has resulted once they have been brought to completion. This is even truer during the research and development (R&D) stages of a novel process, where it is essential to monitor the process of the devulcanisation reaction to ensure that its efficiency is being optimised and that the occurrence of any undesirable side reactions (e.g., those involving chain scission or other forms of main-chain modification) are minimised.

As is usually the case, the need for analysis and characterisation does not stop once a devulcanisation process has been fully developed and commercialised, as it is important to carry out quality control checks at regular intervals as an integral part of any quality assurance and good manufacturing practice system. Carrying out such tests provides the purchasers of the devulcanised product with reassurance that the system is not only inherently capable of producing a high-quality product, but that it is also able to do this consistently over an extended period of time.

It is also the case that once devulcanised rubber has been incorporated into a rubber matrix, or is used on its own, testing is required to ensure that it has the appropriate processing characteristics, curing characteristics and, once vulcanisates have been generated, physical properties for the intended application. Mention of these types of evaluations was a common occurrence throughout the sections in Chapter 4 that reviewed the different devulcanisation processes as they demonstrate if the ultimate goal has been reached; i.e., does the devulcanised rubber, high quality though it might be in terms of a large reduction in crosslink density without loss of molecular weight (Mw), actually do the job that is needed throughout the various processing stages and in service?

The types of processing and physical property tests that are required to assess the important characteristics referred to above are also continually cited in the later sections of this book dealing with the use of waste rubber crumb in rubber products, thermoplastics, and thermosets, for the same reasons (Chapter 7). Many other specific property tests (e.g., acoustic) are also referred to in this book but, unfortunately, there is insufficient space here to cover them. A reasonably detailed section on the characterisation of rubber crumb is provided in Chapter 6, Section 6.4, because understanding the nature of this material is as important as understanding the properties of devulcanised rubber when it comes to its re-use applications.

https://doi.org/10.1515/9783110644142-005

In order to provide these characterisation data it is possible to employ both chemical analysis techniques and physical testing methods. The background to these types of techniques and their application to rubber materials and products have been the subject of a number of texts, including those by Forrest [1], Loadman [2] and Brown [3] and it is not appropriate here to describe the techniques in detail. Instead, a concise summary is provided in Sections 5.1.1 to 5.1.4 of the types of tests that workers in this area have found to be the most useful in their studies.

5.1.1 Chemical analysis tests that can be carried out on waste rubber prior to devulcanisation

It can often be necessary to carry out some initial compositional characterisation tests on samples of waste rubber prior to carrying out any devulcanisation work. The reasons for this could vary from a quality control function, as dictated by a quality system linked to good manufacturing practice, to the need to obtain fundamental compositional information (e.g., rubber type(s), proportions of major constituents, type of filler(s) and so on) in order to carry out some R&D work on a new devulcanisation process.

The chemical analysis tests that are carried out to obtain this type of information tend to only use relatively small amounts of samples (e.g., 10 mg to 1 g) and can be conducted on the rubber whether it is in the form of complete products (e.g., tyre tread, weather strip and so on), crumb or powder.

Two of the most common types of tests that are carried out on rubber samples are those that ascertain the type of polymer that a rubber has been manufactured from and those that determine its bulk composition.

To determine the polymer(s) that are present in a rubber it is usual to use either Fourier-transform infrared spectroscopy or nuclear magnetic resonance (NMR) spectroscopy. These techniques are normally applied to a sample that has undergone some preliminary preparation. For example, a solvent extraction can be carried out to remove process oils and other low-Mw organic compounds and then the extracted sample is often pyrolysed to remove any interferences from the fillers present. As solid-state NMR techniques become more advanced and sensitive, however, it may become easier to obtain this information directly on samples in the future.

With regard to the determination of the bulk composition of the rubber, this is usually achieved by thermogravimetric analysis. This method only requires a small sample (e.g., 10 mg) and is capable of providing a quantitative determination of the amount of process oil, polymer, carbon black and inorganic constituents in the rubber sample. This type of information can also be obtained on larger samples (e.g., 5 g) by combining two or more techniques, for example, by carrying out a quantitative solvent extraction and then heating the extracted sample in a tube furnace under first nitrogen and then an oxygen atmosphere.

If further information is required on a rubber sample (e.g., type of process oil, inorganic filler or cure system) then additional analysis work will need to be carried out using either spectroscopic, thermal or chromatographic techniques, or combinations thereof.

5.1.2 Tests that can be carried out to assess the quality of devulcanised rubber

To determine the effectiveness of the devulcanisation process in removing crosslinks, the sol fraction (i.e., fraction that is soluble in a solvent) can be determined using a solvent extraction test with a solvent that has a high affinity for the particular rubber that is being examined, e.g., chloroform or acetone in the case of natural rubber (NR) vulcanisates. The greater the sol fraction the more crosslinks the process has removed. An important point to make here is that if degradation of the rubber, leading to chain scission, has occurred as a result of the process, this will also increase the sol fraction and so the result that is obtained could be misleading. In order to ensure that the increase in sol fraction is due to crosslink destruction alone, the Mw of the sol fraction can be determined using the technique of gel permeation chromatography.

It is also possible to determine the relative degree of crosslink removal by equilibrium swelling using a high-affinity solvent. The greater the volume swell of a particularly sample, the lower the crosslink density of the matrix.

Physical tests that directly measure the viscosity of a rubber compound can also be useful. A good example of such a test is the one that can be conducted using a rubber process analyser to determine the modulus of a rubber compound. As the crosslinks are removed from a rubber the modulus will be reduced accordingly. If 'standards' are prepared by vulcanising samples to various degrees of crosslink density to obtain calibration data of crosslink density against modulus, the amount of crosslink destruction from a fully cured rubber can be estimated. Care has to be taken, of course, to ensure that other factors that affect modulus (e.g., proportions of the major constituents – polymer, filler and process aids) are the same, and so the calibration approach can only provide accurate data when the composition of the rubber that is being devulcanised is understood and does not vary too much. This can be a problem with large-tonnage waste products, such as tyre tread crumb that has been generated from a range of different types and brands of tyre.

An instrument that functions in a similar way to a rubber analyser is the Mooney viscometer, and this can also be used to measure and assess the extent of devulcanisation.

5.1.3 Tests that can be carried out to determine processing properties of devulcanised rubber

Once a rubber has been devulcanised to the desired level, with the Mw retained at the target level, it has to have cure characteristics (e.g., scorch time) that ensure it is processible. This is particularly true if the devulcanised rubber is to be used with little, or no, modification, but also important if it is to be blended into another rubber compound, although the amount that is to be incorporated into the blend obviously has an impact on the significance of this property.

The scorch time and cure time at a given temperature of the devulcanised rubber can be determined using a rheometer (e.g., a Monsanto rheometer).

When it comes to assessing the quality of the product that can be obtained from the devulcanised rubber during shaping operations (e.g., extrusion) a laboratory extruder fitted with a die that has been developed for this purpose (e.g., a Garvy die) can be employed. This is a small-scale test that prevents large amounts of product being consumed by a full-scale trial. If the results from this laboratory trial look promising, then the full trial can be carried out with much greater confidence.

Other processing trials can be carried out, preferably under representative industry conditions, to assess the performance of the devulcanised rubber during compression moulding, injection moulding, transfer moulding or any other technique used to produce the final product.

5.1.4 Tests that can be carried out to obtain physical properties of vulcanisates containing devulcanised rubber

Once the vulcanisates have been produced, their performance needs to be fully evaluated against any in-house, industry or national/ international standards that are applicable. Some of this work will be carried out on final products (e.g., seals) and examples of these tests are provided in Table 5.1 in Section 5.2.2.1.

However, many tests, particularly the fundamental property tests that are used to acquire physical data such as tensile strength, tear strength, oil and fluid resistance, and ageing performance are usually carried out on specially prepared test sheets due to the need for standard test pieces of a particular geometry.

The results obtained by this suite of tests will enable a final decision to be made, earlier decisions having been made during the progress that has been made up to this point in the recycling process, as to whether the devulcanised rubber is capable of being used for the intended application.

Table 5.1: Commercial products manufactured using tyre tread rubber devulcanised using the DevulCO$_2$ process.

Commercial product manufactured	Key property measured to compare quality with commercial product
Golf tees	Impact resistance
Nozzle protectors	Abrasion resistance
Ground box seal	Tensile properties
Crown pads	Tensile properties, tear strength and hardness
Ladder feet	Wear, marking and slip resistance
Offshore clamp pad	Compression set
Pipe gasket	Compressive creep and shear
Boot seals	Tensile properties, tear strength, compression set, hardness
Anti-vibration mount	Load and deflection
Instrument boot	Impact resistance
Sealing rings	Hardness and compression set

5.2 Characterisation of products manufactured from devulcanised rubber

5.2.1 Introduction

This section includes results obtained by studies carried out on samples that have been produced using rubber that has been devulcanised. These samples have been produced either from devulcanised rubber that has been revulcanised on its own, or from blends prepared by blending the devulcanised rubber with virgin rubber.

It is important to stipulate therefore that this section covers a different type of recycled product than Chapter 7, which reviews the work that has been carried out to produce and characterise products from blends of rubber crumb with virgin rubber, thermoplastics and other materials (e.g., asphalt). In Chapter 7, the rubber crumb that is used is either unmodified, or has been through a surface activation process, but it has not been devulcanised.

During the course of the R&D work that is carried out to perfect a new devulcanisation process, one of the logical steps after the initial characterisation work to assess the quality of the devulcanised product (Section 5.1.2) is to incorporate it into a rubber compound and determine the properties of the resulting material (Sections 5.1.3 and 5.1.4). This is usually done by blending the novel devulcanised rubber with virgin rubber at known levels of addition, vulcanising the blend, and then comparing the data obtained (e.g., physical properties) with those from a control sample (i.e., 100% virgin rubber).

A large number of the results that have been obtained in this way, or recorded on samples to demonstrate the effectiveness of a particular process for commercial reasons (e.g., marketing), have been left embedded in the various sections that make up Chapter 4 if it was thought appropriate to do so in the context of describing a particular devulcanisation process. They were also left embedded if the amount of information that was available was thought insufficient to justify it being removed from Chapter 4 and placed in this one. This section tends therefore to include results from the more extensive characterisation studies that have been carried out on individual processes, or from sources such as comparative studies of a number of processes, and research projects.

One of the main reasons for placing this type of information in a separate section within the book was to keep Chapter 4 reasonably balanced and to not overload certain sections within that chapter with information, particularly Section 4.5. It was also thought that it was more appropriate to place this information within a 'characterisation chapter' rather than a chapter that was focused on describing different devulcanisation processes. So, the information in this section is intended to complement the information provided in Chapter 4, and inclusion in this section does not imply any special merit to a particular process or study, but is often a consequence of the amount of information that is available in the public domain.

5.2.2 Published results

This section has been divided up, for convenience, into studies that have reported on the results obtained from a single process and comparative studies that have compared the products that have been generated by two or more different devulcanisation processes.

5.2.2.1 DevulCO$_2$ and ReMould research projects

The research carried out by two consortia led by Smithers Rapra during two successive devulcanisation projects has been introduced in Chapter 4, Section 4.5. This section provides a fuller description of the projects and the research that was carried out. The first project, DevulCO$_2$, ran from November 2006 to April 2009, had a budget of around £1 million from UK TSB, and had six UK companies in its consortium.

The objective of this project was to develop a novel process to devulcanise waste tyre rubber for manufacturing high-quality products, and its target market was the general rubber goods (GRG) sector. The technology developed during the project, called the DevulCO$_2$ technology, was successful, and a UK patent [4] was published on 21st December 2011. The process described within this patent falls within the 'mechanical devulcanisation assisted by chemical devulcanising agents' category and used waste rubber crumb (20–60 mesh) from truck tyre tread as its raw material. The devulcanisation step involved the use of extruders (either twin-screw or single-screw types) at a relatively low temperature and is assisted by the use of liquid CO_2.

The second research project, ReMould, ran from October 2009 until June 2012 and had a budget of ~£1 million from the UK TSB. Its aims were to build on the achievements of the DevulCO_2 project by carrying out the following:

- Achieve commercial scale-up of the DevulCO_2 technology.
- Transfer of DevulCO_2 technology from tyre rubber to sulfur- cured GRG products, e.g., nitrile rubber matting and ethylene- propylene-diene monomer (EPDM) rubber automotive weather strip.
- Improve processibility to enable extrusion profile work, e.g., manufacture of recycled weather strip.
- Retreading of tyres using relatively high levels (e.g., 30%) of DevulCO_2 rubber.

This project consortium had eight members: the six members of the DevulCO_2 project and two additional members. During the course of the two projects, a number of practical studies were carried out to assess the performance of the DevulCO_2 rubber (derived from waste tyre tread and from GRG products); some of these results are described in a paper by Forrest [5].

The work carried out during the DevulCO_2 project showed that the devulcanised waste tyre tread rubber could be successfully processed using a mill, calender or extruder, and could be revulcanised into good quality products using compression moulding, transfer moulding or injection moulding. By varying the recipe, three categories of DevulCO_2 products could be marketed; these are shown below with an example of how they could be used by industry:

- Masterbatch (MB) (e.g., for blending in intermixers with virgin compounds)
- Sheetable product (e.g., from a two-roll mill for the preparation of moulding blanks)
- Extrudable product (e.g., for feeding injection-moulding machines)

A wide range of commercial products were produced from the tyre tread rubber that had been devulcanised using the DevulCO_2 process by members of the consortium (Figure 4.3). The key physical properties of these products were determined to see how they compared with commercial products produced using virgin rubber. The commercial products manufactured from the DevulCO_2 rubber and the key properties that were determined to assess their performance are shown in Table 5.1.

The conclusions from the results obtained for these products were that the DevulCO_2 rubber could be substituted for virgin rubber in the manufacture of a number of them, but that there were potential problems when good tear strength was required, or where the specification was very high. In the former case, the addition of a filler such as silica was found to improve the tear strength, and in the latter case, blending with virgin rubber is a possibility.

Another investigation undertaken during this project was concerned with the evaluation of the effects of recycling the DevulCO_2 devulcanised rubber itself. To carry this out, a number of Crown Pad products were produced using the DevulCO_2 rubber and then were ground up into crumb. This crumb was then subjected to the DevulCO_2 process and data obtained from a test sheet and compared to the original DevulCO_2 values. The results showed that a loss of around 50% occurred in the tensile strength, but that the values for hardness, tear strength, and elongation at break were similar.

At the conclusion of the project, the benefits of the DevulCO_2 process for devulcanising waste tyre rubber could be summarised as follows:

- Uses standard equipment with some modification
- No additional health and safety or environmental concerns
- Devulcanised rubber has good processing characteristics
- Devulcanised rubber can be used to produce high-quality products for the GRG sector

The second research project, ReMould, was also successful in achieving its goals. Scale-up of the DevulCO_2 technology was achieved using an extruder, which demonstrated that it had the capability of manufacturing large quantities of good-quality devulcanised rubber. Regarding the other targets, the results showed that the DevulCO_2 devulcanisation technology could be successfully transferred from waste tyre rubber to a number of other sulfur-cured rubbers, including nitrile rubber, EPDM rubber and NR. For extruded weather strip, in addition to a satisfactory result in terms of the degree of devulcanisation achieved, the other property that was crucial was surface finish. To achieve the quality of extrudate that is required by industry it was found to be necessary to blend the DevulCO_2 rubber with virgin EPDM compound, e.g., 20% DevulCO_2/80% virgin EPDM. This demonstrated that it was possible to use the DevulCO_2 to return relatively large amounts of waste EPDM weather strip into the production stream and so assist manufacturers in achieving the targets set by the End-of-Life Vehicle Directive.

The other major objective of the ReMould project was to evaluate the performance of the DevulCO_2 rubber in tyre retread compounds. To achieve this, a number of passenger tyres (size 185/65) were retreaded using tread compound blends that contained from 5 to 30% of DevulCO_2 devulcanised tyre rubber (Figures 5.1–5.3). Although only a limited number of tests could be performed within the scope of the project, the results were very encouraging. For example, a range of six tyres containing from 5 to 30% DevulCO_2 rubber in their tread all passed the requirements of the standard ECE Regulation 108 Drum Test. This test measures heat build-up, pressure changes and dimensional changes. In addition, rolling resistance and force and moment tests were also conducted. Rolling resistance measurements are taken as a means of measuring fuel economy. This test was performed on tyres that had

Figure 5.1: Passenger tyre carcasses retreaded with compounds containing 5–30% DevulCO_2 rubber prior to moulding. Reproduced with permission from the DevulCO_2 Research Project Consortium (Lead Partner: Smithers Rapra, Shawbury, UK), UK. ©2012, DevulCO_2.

from 5 to 30% DevulCO_2 rubber in their tread compound, and the results obtained showed that, although the rolling resistance varied over the range, there was no discernible increase in rolling resistance as the amount of DevulCO_2 increased and it was thought that any variation that was apparent was due to the variation in the weight of the retread tyre carcasses. Tyre force and moment measurements are used to identify tyre performance characteristics and provide an insight into how a tyre's design affects overall vehicle dynamics. The specific tests that were carried out on the 5–30% DevulCO_2 tyres were cornering stiffness, cornering force and weight loss. As the amount of DevulCO_2 rubber in the tread increased, the cornering stiffness results showed a general improvement, the cornering force was found to reduce, but there was no apparent trend either way with respect to weight loss.

(a)

(b)

Figure 5.2: Passenger tyre retreaded with DevulCO_2 rubber in the mould prior to be being moulded a) and after moulding b). Reproduced with permission from the DevulCO_2 Research Project Consortium (Lead Partner: Smithers Rapra, Shawbury, UK), UK. ©2012, DevulCO_2.

5.2.2.2 DeLink

Some of the earliest cited practical, third-party trials of DeLink were reported in *Plastics and Rubber Weekly* [6]. The article covered the official launch for DeLink at the 'International Seminar and Workshop on Devulcanisation Using DeLink R Process' that was held at the TARRC laboratories (then MRPRA). Papers that were presented generally endorsed the claims made by STI-K that the rubber devulcanised

Figure 5.3: Set of cured passenger tyres retreaded with DevulCO_2 compounds waiting to be trimmed. Reproduced with permission from the DevulCO_2 Research Project Consortium (Lead Partner: Smithers Rapra, Shawbury, UK), UK. ©2012, DevulCO_2.

by DeLink can be used in manufacture or blended with fresh rubber compound to obtain the desired levels of technological performance. Rubber that had been devulcanised with DeLink had been used successfully to:

- Produce an alternative tyre tread material (from used tyre tread buffings).
- Produce injection-mouldable thermoplastic materials when blended with fresh NR (e.g., Standard Malaysian Rubber grade L) and polypropylene (PP).
- Demonstrate good static and dynamic fatigue and resistance to reversion (from tyre crumb).

This seminar was accompanied by press releases for DeLink in *Rubber and Plastics News* [7] and *The Business* [8] and papers presented by STI-K personnel at international conferences, e.g., Roland Newell [9] and Hon and co-workers [10]. The latter provided an overview of the DeLink technology and provided practical recommendations as to how it should be used to devulcanise waste rubber and how different processing parameters affected the quality of the resulting material. This paper also provided a number of examples of rubber formulations where the De-Vulc devulcanised material had been blended with virgin styrene-butadiene rubber (SBR)/butadiene rubber (BR) tread rubber, the ratio of the two materials varying up to a maximum level of 70% De-Vulc to 30% SBR/BR, as well as the properties that had been obtained on the resulting vulcanisates. In addition to being used to manufacture new rubber articles (e.g., retread tyres, gloves, shoe soles and rubber sheet), the De-Vulc rubber was also reported as being capable of being used to produce thermoplastic rubbers in blends with PP and NR, a blend of De-Vulc 50%, PP 30% and NR 20% being described. A comparison of a 100% De-Vulc NR-based tyre tread and the virgin material was provided and is reproduced in Table 5.2.

Table 5.2: Comparison of 100% De-Vulc NR-based tyre tread and the virgin material [10].

Property	Virgin	100% De-Vulc	Recovery
Cure time at 140 °C (min)	23	25	+2
Tensile strength (MPa)	29.0	19.7	68%
Elongation at break (%)	533.7	275.6	52%
Modulus at 100% (MPa)	2.51	3.66	146%
Hardness (international rubber hardness degrees)	65.0	72.0	+7

A paper was published by Ishiaku and co-workers [11], who investigated the optimum concentration of DeLink to use in the devulcanisation of a sulfur-cured, NR-based powder that had originated from waste rubber balls and artificial eggs. A solubility test was also developed to assess the degree of crosslink destruction that had taken place. The results obtained showed that the optimum concentration of DeLink was 6 phr (which agreed with the manufacturer's recommendation), with the devulcanised product never becoming fully soluble and the strength of the vulcanisate being dependent upon a number of factors including the level of DeLink and the amount of degradation that occurs due to the influence of mechanical shear and heat reversion. The range of values for the tensile strength and elongation at break of the revulcanised rubber were 2.0–2.5 MPa and 380–440%, respectively. The original rubber had a tensile strength of around 7.5 MPa and an elongation at break of 680%. In this set of experiments, the DeLink therefore struggled to retain the 50–80% of the original properties of the original rubber as indicated by its developers.

Gilannejad and Hepburn [12] have described the use of 2–3 mm tyre crumb that has been devulcanised by DeLink for use as a functionally active component for the adhesion enhancement of low-cost plasma-activated vulcanisates. The devulcanised rubber was mixed with a fresh silica-filled NR compound, and at high levels (60% of the total compound) was found to both increase the tensile strength (from 12 to 15 MPa) and enhance its bonding properties when its surface was activated by plasma. However, the workers also found that the devulcanised rubber was not completely soluble in trichloroethylene, a thick high-solids (20–30%) dispersion resulting, and therefore concluded that complete devulcanisation could not have occurred. They also found that the tensile strength of the pure revulcanised, DeLink-treated tyre tread was only around 6 MPa.

A more recent article that addressed the use of the DeLink process was published in *Rubber Journal Asia* [13]. It featured the Gujarat Reclaim and Rubber Products Company, which produces reclaimed rubber from both synthetic and NR waste, as well as a light-coloured reclaim from latex scrap using the Rubplast process. The company uses Green Rubber's patented devulcanisation agent, DeLink, at a ratio of two parts DeLink to 100 parts crumbed waste rubber, to produce a product that can be incorporated back into new rubber compounds. The article claims that this is very advantageous to the rubber industry as it enables them to re-use the 5–15% of waste, which they typically throw away.

DeLink recycled rubber has been used successfully in the manufacture of commercial shoe soles. In 2009 a news story was released on the Fox Business website concerning the use by the Timberland Company of recycled rubber in the outsoles of its shoes [14]. These outsoles would be manufactured using a blend that consisted of 42% Green Rubber material (produced using their DeLink process) and 58% virgin rubber. This blend could be used without compromising the durability and performance characteristics of the footwear. The article, which featured quotes from Timberland and Green Rubber representatives, forecast that Timberland would use 50 tonnes of Green Rubber material in the autumn of 2009 in the outsoles of more than 200,000 shoes, which would replace 42 tonnes of virgin rubber, and that by 2012 Green Rubber intended to recycle the equivalent of more than 200 million discarded tyres every year.

5.2.2.3 Levgum

The Levgum process has been described in Chapter 4, Section 4.5, and the company have published on their website some data obtained from vulcanisates produced from blends of tyre buffings that were devulcanised using their technology (designated devulcanised rubber compound, DRC) and a MB (containing fillers and so on) based on a fresh NR compound. A range of ratios of DRC to MB have been used, e.g., MB/DRC = 85/15, 75/25, 50/50 and 30/70. After blending the DRC and MB, a sulfur-based cure system was incorporated into the resulting mix. The tensile strength of these compounds varied from 257 to 224 kg/cm^2, and the elongation at break from 538% to 387%. The T90 rheometer cure times at 160 °C varied from 4.66 to 2.05 min, and the scorch times from 1.58 to 0.65 min. Together, these data show that increasing the amount of DRC reduced both the physical properties and the cure times/scorch times of the mixes. Levgum have explained [15] that incorporating 25% of the DRC in a virgin compound results in a reduction of physical properties of less than 10%, and that using a 50% level can result in around a 20% reduction.

5.2.2.4 Rubber resources (formerly called vredestein rubber recycling)

Antonio [16] has demonstrated how the final properties of a revulcanisate can vary according to a combination of the type of scrap rubber that is used as the starting material and the devulcanisation conditions that are used to devulcanise it. The study demonstrated this by using four NR compounds with tensile strengths varying from 4.4 to 27.9 MPa, generating crumb from them and then devulcanising the crumb using an essentially two-stage process that initially heated the rubber (usually with a devulcanisation agent) *via* frictional forces and then subjected it to high shearing forces in an internal mixer. The resulting devulcanised rubber was compounded with a sulfur- based cure system (mercaptobenzothiazole and diphenyl guanidine (DPG) as accelerators) and vulcanised in a press at 150 °C. The resulting vulcanisates were found to have tensile strengths between 3.5 and 13.2 MPa.

In another paper, Wijers [17] described the optimisation of the devulcanisation technology with a view to improving the physical properties of NR-based revulcanisates. Again, a two-step process is described. In the first stage, truck tyre tread peelings are mixed with a devulcanisation agent and then heated in a specially designed paddle stirred reactor up to the required reaction temperature. The rubber/agent mix is then exposed to high frictional forces. The paper concentrates on the effect of adding a 'heterogeneous catalyst' to reduce the cycle time and process temperature, the effect of which is to improve the tensile strength and elongation of the revulcanised products that are produced from the reclaimed rubber, probably due to a reduction in the amount of degradation/chain scission that occurs. One adverse effect of the catalyst was to increase the Mooney viscosity of the devulcanised rubber. The tensile strength of the catalysed product (after revulcanisation) was around 10 MPa, whereas the conventionally devulcanised rubber was around 8 MPa; the original, virgin material had a tensile strength of 18.8 MPa. No information on the identity of the original devulcanisation agent, nor of the catalyst, are provided in the paper.

5.2.2.5 Miscellaneous processes

Sangari and co-workers [18, 19] have used a two-litre Banbury intermixer and a combination of accelerators with a peptiser to devulcanise both standard SBR compounds (unfilled and filled) and ground tyre tread. The devulcanisation recipes that were used in the study are shown in Table 5.3.

Once devulcanised in the Banbury using a 3.5 min (0.5 min initially, then 3 min with the mix of chemical agents) programme, the devulcanised samples were mixed with 0.4 phr cyclohexyl benzothiazole (CBS), 5 phr zinc oxide and 1 phr stearic acid and revulcanised at 160 °C. The time that was used for each material

Table 5.3: Examples of devulcanisation recipes used by Sangari and co-workers [18].

Ingredient	Quantity (phr)	Quantity (phr)	Quantity (phr)
Rubber[a]	100	100	100
MBTS[b]	0.68	0.68	0.68
CBS[b]	0.08	0.08	0.08
DPG[b]	0.15	0.15	0.15
Peptiser	0.5	2.00	4.00

[a]SBR compound or tyre tread rubber
[b]These accelerators (ratio: MBTS 0.65, CBS 0.48 and DPG 0.15) had also been used to vulcanise the original SBR rubber compound (along with 2 phr sulfur, 5 phr zinc oxide and 1 phr stearic acid) that was being devulcanised in these experiments
CBS: Cyclohexyl benzothiazole
MBTS: Mercaptobenzothiazole disulfide

was obtained from the T90 (i.e., 90% cure state) time value recorded in a rheometer trace.

The workers found from the results obtained on the SBR compounds that the presence of carbon black can reduce the overall degree of crosslink density as it acts as a free radical scavenger and stops new crosslinks forming during the devulcanisation reaction as a result of the generation of free radicals. The peptiser added to the mixes also had this effect. In the case of the tyre tread, the revulcanised rubber was found to have a tensile strength of around 10 MPa and an elongation at break of 200%, which compares with 19 MPa and 250% for the unused, virgin tread compound.

A standard production thermo-mechanical 'reclaiming machine', designated MRR-031, has been developed by Elast-Tekhologii LLC in Russia [20, 21]. This machine is capable of reclaiming rubber from scrap tyres and GRG based on SBR, polyisoprene, EPDM and butyl rubber. The reclaimed rubber has been incorporated back into new rubber compounds at a replacement level of up to 90% of virgin rubber polymer. The physical properties that are reported for various types of rubber compound containing high levels of devulcanised rubber are shown in Table 5.4.

Table 5.4: Physical property data reported by Elast-Tekhologii LLC [20, 21].

Compound	Tensile strength (MPa)	Elongation at break (%)
SBR	7.7	160
Polyisoprene	5.0	130
EPDM	7.3	180
Butyl rubber	5.0	200

Overall, the property values in Table 5.4 are relatively low and less than those reported using other methods of devulcanisation.

Workers at Sichuan University in China [22] have carried out extensive characterisation work using a range of analytical techniques on three different samples:

- Devulcanised ground tyre rubber that resulted from a mechano- chemical process.
- Vulcanisates that resulted from blends of the devulcanised ground tyre rubber with virgin rubber.
- Vulcanisates that resulted from blends of the original, untreated ground tyre rubber with virgin rubber.

In the case of the devulcanised rubber, properties such as crosslink density, extractable fraction and gel fraction were obtained. For the two different series of vulcanisates, cure characteristics, and morphological and mechanical properties were among those that were investigated.

A team in Thailand [23] have investigated the effect of recycling on blends of chlorinated polyethylene (PE) and NR vulcanised using a sulfur-based cure system. A number of different blends were prepared and the viscoelastic properties of them determined as a function of the recycling cycle. The influence of magnesium oxide as an acid acceptor for chlorinated PE on the viscoelasticity of the blends was also investigated. The changes in the viscoelasticity that the workers found during recycling were explained in terms of thermal degradation of the material *via* a molecular chain scission mechanism.

Workers in Sri Lanka [24] reclaimed waste rubber by utilising a readily available, low-cost and environmentally friendly amino compound and then carried out an extensive study comparing the properties of a compound produced by blending virgin NR with this reclaimed rubber at a ratio of 85:15 (designated Novel reclaim) with two other rubber compounds:

- A compound produced by blending virgin NR with standard reclaim rubber at a ratio of 85:15 (designated the Control reclaim).
- A compound that was 100% virgin NR (designated as the Control NR).

Samples were cured from each of the three compounds and then a large number of physical and chemical properties were examined, including tensile properties, tear strength, resilience, abrasion resistance, ageing, and performance and swelling in test fluids. The results revealed that the abrasion volume of blends of virgin NR and the Novel reclaim is lower than that of the Control reclaim and comparable to the Control NR. Although the tensile properties of the Novel reclaim were found to be lower than Control NR, the workers concluded that they would be acceptable for a tyre tread application and overall it was concluded that the 85:15 Novel reclaim product could be used as an alternative to both the other two compounds for tyre tread manufacture.

5.2.2.6 Comparative studies

In a report published in July 2007, the Waste Resources Action Programme (WRAP) published the results of a research project that they had funded at Tun Abdul Razak Research Centre (TARRC) in the UK [25]. In this project, WRAP worked with a supplier of waste tyre crumb, J. Allcock & Sons, and a research institute (TARRC) to develop improved chemical systems that would lead to a commercially viable process for recycling end-of-life tyre rubber at high levels into high-value new rubber products with demanding specifications. The aim of the project was to develop improved chemical systems for efficiently breaking the sulfur crosslinks in waste tyre rubber. A further aim was that these novel systems, for both devulcanisation and revulcanisation, must be at least as practical and convenient as those currently used by the rubber industry, including a reasonable level of safety with regard to 'scorch'. As some existing systems suffer from these types of problems, it was hoped that if these aims could be achieved, the level of exploitation of devulcanisation as a recycling route within the rubber industry would be significantly increased. The project assessed the performance of a number of types of devulcanising systems, which could be arranged into three groups:

- Combinations of thiophilic chemicals currently used by the rubber industry that fall within, or are close to, the scope of the patents covering the DeLink system.
- Alternative thiophilic compounds that are not currently used as rubber chemicals but that were anticipated to behave in a similar manner to the DeLink chemicals.
- Chemicals that are used in the Levgum process and those that may work in a similar manner to this particular devulcanisation technology.

The scope of work that was carried out included:

- Devulcanisation of waste tyre crumb (40 or 45 mesh) with the chemicals using a standard procedure based on the DeLink literature.
- Determining the scorch safety of the samples (devulcanised rubber alone and in blends with virgin rubber) using rheometry and Mooney scorch delays.
- Evaluating the devulcanisation by Mooney viscosity and Horikx analysis.
- Evaluating the revulcanised samples (devulcanised rubber alone and in blends with virgin rubber) using rheology and various physical tests.

At the conclusion of the work, the project was regarded as having achieved its primary target of formulating new systems to improve the scorch performance that could be obtained by a commercial DeLink system by using combinations of chemicals that were less active at the revulcanisation stage. However, the workers found that no single system could be used for all applications as each had its limitations. There were two principal conclusions:

- The chemicals that were more efficient at devulcanisation gave the least scorch safety.
- These chemicals were also found to produce better physical properties when the devulcanised rubber was incorporated into new products.

The project's goals, to achieve both an improvement in scorch performance and good physical properties, were not fully achieved, but WRAP felt that there was potential for them to be achieved in the future, for example by using a combination of a high level of a less hindered sulfenamide, such as CBS, with a small amount of a less active zinc-based accelerator, such as zinc ethylphenyl dithiocarbamate.

The paper presented by Kind at *RubberChem 2010* [26] described the processes that the UK company Polymer Recyclers Ltd (PRL) had developed to convert powdered end-of-life rubbers into high- quality rubber ingredients that are in the form of MB-containing small amounts of virgin rubber. PRL has given these MB the trade name ReTurn. The presentation also provided a technical comparison of the results obtained using ReTurn and those obtained using the commercial alternatives Levgum and DeLink. For these studies, the starting material was 40 mesh truck tyre rubber and, with the use of a laboratory mill, samples were prepared from it using PRL's own patented ReTurn process, as well as some Levgum samples using the information provided in the Levgum literature, and some DeLink samples using the DeLink system under the recommended conditions. All three materials were then added to a NR base with a cure system added to make compounds that had a recycled rubber content of 60%. A Control sample was also prepared using 40 mesh crumb rubber that had only been milled. These compounds were then tested for viscosity and vulcanisation characteristics and, once vulcanised, physical properties. Some of the results are shown in Table 5.5.

Table 5.5: Comparison of the results obtained using DeLink, Levgum and ReTurn technologies with a control sample prepared using 40 mesh unmodified crumb rubber.

Property	Control	DeLink	Levgum	ReTurn
Mooney ML 1 + 4 100 °C	77	50	42	63
ODR at 140 °C scorch (min)	2.92	1	1.93	2.20
ODR at 140 °C 90% cure (min)	18.60	4	16.30	14
Tensile strength at break (MPa)	11	11.80	15.30	12.40
Elongation at break (%)	307	225	302	311
Trouser tear (N/mm)	5	7.40	12	12.70
DIN abrasion (% volume loss)	5.80	6.60	5.85	5.25

DIN: Deutsches Institut für Normung e.V.
ODR: Oscillating disc rheometer

Kind was able to conclude from these data that only milling crumb without the addition of chemicals was ineffective and that the Levgum and ReTurn processes gave good tensile, tear and abrasion loss results, with similar values. It was stressed that it was difficult to translate the obtained results into results that would be obtained commercially as the milling process was carried out on a relatively small scale.

In addition to the comparison study, the paper also described the results of a number of case studies. In these, the ReTurn process was applied to different sources of waste rubber and used, at various levels of addition, to produce samples of the following compounds:
- Truck tyre retreads
- Nitrile calendered sheet
- EPDM drainage seals
- Butyl curing bladders
- Butyl tyre inner liners

A number of properties were assessed in all cases and good results were obtained. One of the conclusions reached from the work was that formulations using high proportions of recycled vulcanised rubber can produce good properties, but that the formulations may need to be designed to improve processability. The results also showed that end-of-life rubber does not need to be devulcanised for it to be returned to a useful product, and that a 'closed loop' scenario was possible in the case of industrial scrap.

References

1. M.J. Forrest in *Rubber Analysis – Polymers, Compounds and Products*, Rapra Review Report No.139, Rapra Technology, Shawbury, UK, 2001.
2. M.J.R. Loadman in *Analysis of Rubber and Rubber-like Polymers*, 4th Edition, Kluwer Academic Publishers, Dordrecht, The Netherlands, 1998.
3. R. Brown in *Physical Testing of Rubber*, Chapman and Hall, London, UK, 1996.
4. M.J. Forrest and L. William Lloyd, inventors; Smithers Rapra Technology Ltd, assignee; GB 2481205, 2011.
5. M.J. Forrest in *Proceedings of the RubberChem Conference*, Vienna, Austria, Smithers Rapra Technology Ltd, Shawbury, UK, 2010, Paper No.17.
6. *Plastics and Rubber Weekly*, 1995, 12th May, 7.
7. M. Moore, *Rubber and Plastics News*, 1996, 7th October.
8. M. Moore, *The Business*, 1995, 20th February, 10.
9. R. Newell in *Proceedings of Rubber in the Environmental* Age – *Progress in Recyling*, 18th November,Rapra Technology Ltd, Shawbury, UK, 1996, Paper No.7.
10. K-K. Hon, F. Siesseger, B.C. Sekhar and S.W. Sin in *Proceedings of the ACS Rubber Division Meeting – Fall 1995*, Cleveland, OH, American Chemical Society, Washington, DC, USA, 1995, Paper No.55.
11. U.S. Ishiaku, C.S. Chong and H. Ismail, *Polymer Testing*, 1999, **18**, 621.
12. Gilannejad and Hepburn in *Proceedings of the International Rubber Conference*, Kuala Lumpur, Malaysia, International Rubber Conference Organisation, Institute of Materials, Minerals and Mining, London, UK, 1997, p.243.
13. *Plastics and Rubber Asia*, 2009, **24**, 167, 7.
14. *Timberland Teams with Green Rubber™ to Reduce Global Tire Waste*, Fox Business. http://www.Foxbusiness.com.

15. PTA News Bureau, *Polymer and Tyre Asia*, 2011, April/May,66.
16. M.E. Antonio, *Tire Technology International*, 2002, 86.
17. B.G.C.J. Wijers in *Proceedings of International Rubber Conference*, Birmingham, UK, International Rubber Conference Organisation, Institute of Materials, Minerals and Mining, London, UK, 2001, p.380.
18. S.S. Sangari, N. Kao, S.N. Bhattacharya, D. Paval and K. Silva in *Proceedings of the ACS Rubber Division Meeting – Fall 2001*, Cleveland, OH, USA, American Chemical Society, Washington, DC, USA, 2001, Paper No.39.
19. S.S. Sangari, N. Kao, S.N. Bhattacharya and K. Silva in *Proceedings of Annual Technical Conference*, Society of Plastics Engineers, Bethel, CT, USA, 2003, p.1648.
20. S.E Shakhovets, Haddog Buzid and V.V. Bogdanov, *Kauchuk i Rezina*, 2006, **2**, 30.
21. S.E Shakhovets and B.L. Smirnov, *Kauchuk i Rezina*, 2006, **1**, 34.
22. X. Zhang, C. Lu and M. Liang, *Journal of Polymer Research*, 2009, **16**, 4, 411.
23. P. Prasopnatra, P. Saeoui and C. Sirisinha, *Journal of Applied Polymer Science*, 2009, **111**, 2, 1051.
24. D.G. Edirisinghe, M.I.A. De Silva and J.K. Premachandra, *Progress in Rubber, Plastics and Recycling Technology*, 2011, **27**, 3, 161.
25. *Recycling of Tyre Rubber into New Rubber Products through Efficient Devulcanisation*, Project: TYR3-12, Waste Resources Action Programme, Banbury, UK, 2007.
26. R.C. Kind in *Proceedings of RubberChem Conference*, Vienna, Austria, Smithers Rapra Technology, Shawbury, UK, 2010, Paper No.18.

6 Production of rubber crumb

6.1 Introduction

In addition to being the starting point for the majority of the devulcanisation processes discussed in Chapter 4, the production of crumb from waste tyres and end-of-life general rubber goods (GRG) is also an important end in itself to produce a wide range of commercial and technical products that have been used in numerous ways to manufacture new articles, by processes like sintering, or by modifying rubbers, plastics and a range of other materials, e.g., cementatious materials and wood. The manufacture of new products by the use of sintering techniques is covered in Section 6.5 and the use of crumb to manufacture a wide range of products is the subject of Chapter 7.

When it originates from tyres, crumb can be produced from either whole tyres or what remains of the tread when a tyre, usually a truck tyre, is to be retreaded. In the latter case, the starting material is referred to as 'tread buffings', produced when the remnants of the tread are 'buffed off', and the crumb is produced by grinding this material. With whole tyres, it is possible to use complex reducing machines that start by breaking the tyre up using a powerful hammer mill and then subsequent stages within the process will grind, separate and segregate the main constituents within a tyre, i.e., the metal, fabric and rubber.

When tread buffings are to be used as the starting material, they are ground using one of the methods described in Sections 6.2. Using tread buffings as a starting material to generate crumb has an advantage over the whole tyre route, in that the potential for contamination is significantly reduced, as the crumb should not be contaminated by either fabric or metal, as they remain within the untouched tyre carcass. Crumb produced using this route should also be less complex than whole tyre crumb when it comes to the range of rubber types present in it and the types of compounding ingredients present. This is due to the absence of the other parts of the tyre, e.g., the sidewall and, in tube-less tyres, the butyl rubber liner.

By using waste products from the GRG sector, rubber crumb from a number of different products and rubber types is commercially available. This type of crumb is normally designated for sale according to its rubber type and particle size, rather than the particular product from which it originated. This can be because more than one type of product can be used as the source of the crumb, and it is more important for consumers of the product to know the rubber type. Two common examples of crumb from GRG are crumb from ethylene-propylene-diene monomer (EPDM) rubber weather strip that has been recovered from cars, and crumb from nitrile rubber matting used in offices and other commercial buildings. These two sources of waste rubber are attractive due to standardisation of the manufacturing technology, which means that the type of rubber present in them is known, and the

https://doi.org/10.1515/9783110644142-006

fact that, like tyres, they are easily identified and collected, a useful attribute given that the majority of rubber products are black, as it helps to avoid contamination by other types of rubber. The problem of accurate and effective separation at source means that a large number of end-of-life GRG products (e.g., general seals, gaskets and diaphragms in cars) will be mixed together, resulting in a range of rubber types, cure systems and compound ingredients being present. This is currently a major problem for the recycling of this type of waste into new products. It is often the case, therefore, that this rubber either has to be recycled in other ways (e.g., energy generation) or placed into landfill.

Irrespective of whether the source of waste rubber is whole tyres, tread buffings or GRG, the crumbing operation is sometimes carried out under cryogenic conditions (i.e., it is cooled to below around –80 °C) with liquid nitrogen. In addition to hardening the rubber to make it easier to crumb via the action of mechanical hammers, this also has the additional benefit of reducing any frictional heat build-up and so minimises heat degradation. However, this is just one of the methods that can be used to produce crumb. This cryogenic process and the other main technologies are described in Section 6.2.

Once the rubber crumb has been produced it can be graded (e.g., 40 mesh) by the use of industrial sieving machines (Section 6.3) and characterised as described in Section 6.4. If the rubber crumb is to be produced from whole tyre waste it is important that the fabric and metal are removed as efficiently as possible during the recycling process so that they do not contaminate the crumb. As described in a number of sections in this book (e.g., Section 6.5), it only takes a small quantity of these types of contaminants to impair the final properties of a product, particularly when it is a rubber product, as they provide a stress concentration point in the material, resulting in premature tearing, breaking and cracking when it is placed under stress.

There are a number of physical and chemical tests that can be performed on rubber crumb in order to characterise it and the results obtained can provide guidance on which products it may be suitable to be used in. For example, Massarotto and a group of Brazilian workers [1] ground up waste crosslinked styrene-butadiene rubber (SBR) foam rubber from the shoe industry using an ambient grinding process and then characterised the resulting crumb using scanning electron microscopy (SEM) and thermogravimetric analysis (TGA) and determined its gel content by a solvent extraction method. The results obtained from these tests enabled the workers to select a crumb that was suitable for use in microcellular composites.

Another example of applying characterisation tests in this way concerns a group in Brazil [2], who started with scraps generated during the extrusion of SBR and EPDM profiles, and ground them under ambient conditions. The rubber crumb materials that were obtained had irregular-shaped particles with a high surface area and sizes in the range 0.425–6 mm. They found by chemical and thermal analysis that the grinding process had not had an adverse effect on the rubber and that it could therefore be used for a wide range of applications.

Finally, the heat stability and ageing resistance of the crumb is often important, and a study of the thermal degradation kinetics of ground tyre rubber has been performed by Gisbert and co-workers [3] using TGA. They characterised the bulk composition of the rubber (i.e., amount of volatile compounds, rubber, carbon black and inorganic ash) and obtained the maximum weight loss rate peaks for the two main polymer components within tyres (i.e., natural rubber (NR) and SBR/butadiene rubber blend) using the Gaussian deconvolution method. In this way they could understand the influence that the particle size of the crumb and the heating rate of the sample had on the reaction rates and degradation time of the ageing process.

Other examples of the types of tests that are mentioned above have been covered in Chapter 5, Section 5.1.1, where they are used to characterise rubber crumb and other waste rubber products to assist in devulcanisation work.

To assist in the marketing of tyre-derived rubber products, a categorisation system has been established as part of the UK Waste Resources Action Programme (WRAP) Quality Protocol (Section 6.4.1.5), which will help avoid possible confusion in the use of terms such as 'rubber crumb' and 'rubber powder'. The categories and definitions that are in the protocol are shown in Table 6.1.

Table 6.1: Definitions in the UK WRAP Quality Protocol for different forms of size-reduced rubber.

Category	Size (mm)
Rough cut shred$_a$	Reduced to a maximum of 300
Clean cut shred$_b$	Reduced to a maximum of 300
Rough cut chips$_a$	Reduced to a maximum of 50
Clean cut chips$_c$	Reduced to a maximum of 50
Granulate$_d$	Reduced to a maximum of 10
Powder$_d$	Reduced to a maximum of 5
Fine powder$_d$	Reduced to a maximum of 0.5

[a]With exposed wire and textiles; can only be used in landfill engineering as replacement aggregate at present

[b]Containing less than 5% exposed wire and textiles on visual inspection

[c]No exposed wire and containing less than 5% exposed textiles on visual inspection

[d]Free from exposed wire and textiles

6.2 Methods for the production of rubber crumb

There are a large number of different techniques whereby waste rubber products and rubber, either vulcanised or unvulcanised, can be reduced in size and ground down into small particles. In general, these processes fall into categories where this commutation in size is carried out using cutting, shearing or impact (or a combination of one or more of these), and can vary in the environment under which this occurs, wet, ambient or cryogenic conditions all being possible.

All these various processes have their associated technical and practical advantages and disadvantages, and the choice of which type to choose is dependent upon a number of factors, such as the starting form of the waste rubber product, the desired morphology, particle size and particle size distribution of the final rubber particles, the degree of residual contamination that can be tolerated, and the desired target price. With respect to price, rubber crumb that is produced from relatively pure sources, such as tyre tread buffings, and in the smaller particle sizes, normally sells for higher prices due to the cost of the buffings and the fact that it is more difficult to produce and can require specialist equipment. However, this type of crumb is also very desirable because of its purity and its relatively high surface area, which usually ensure that the best results are obtained when it is used in devulcanisation processes and when it is incorporated into other materials.

In a research project conducted from December 2005 to February 2006 [4], WRAP sought to provide answers to the following questions. What does it take to make really fine crumb? What is it used for? They provided the funding for Edge Rubber, a small manufacturing company in the Pennsylvanian city of Chambersburg that makes fine rubber powders for use in new tyres, brake linings and pigment, to carry out the work to provide the answers. At the time of the research, Edge Rubber was the only rubber powder manufacturer in the USA to produce a premium, fine-grade product (e.g., 140 mesh) that met demanding standards (e.g., American Society for Testing and Materials (ASTM) D-5644 standard for particle size) using a wet-grind technique. The company also operated an ambient grinding process for the production of high-quality, larger particle-sized powders. The powders that were produced using these two types of processes, from whole passenger tyres, whole truck tyres and retread buffings, were supplied to the following markets:

- The bitumen crack and seal industry (10 and 20 mesh).
- The automotive industry (e.g., tyre manufacturers) (20, 30, 50 and 80 mesh).
- Brake liners (20 and 30 mesh).
- Roofing products and pigments for paint products (e.g., automotive paints) (140 mesh).

The results obtained by this WRAP project highlighted that it is possible to produce high-quality rubber powder and crumb products in a consistent way from waste tyres and that these can be used in a wide range of applications.

6.2.1 Preparation of waste rubber products prior to grinding or crumbing

Sometimes it can be necessary to prepare waste rubber products by reducing their size prior to any grinding or crumbing step. An example of where this is required is when the type of mill that will be used is not sufficiently large enough to be capable of being fed with complete products, e.g., tyres. In such cases, the tyres, and any other rubber product or material, will be cut up first. This reduction process can be done using knives and involves both cutting and shearing of the material. The knives, and other types of cutters, can be present in a wide range of equipment of various designs, e.g., granulators, rotary cutters and shredders. The pieces of rubber that can be generated by these devices range in size from a few centimetres to a couple of millimetres.

Having ensured that the rubber is in a form that is suitable, it is then ground using one of a number of techniques. A brief review of the principal processes that can be used to turn waste rubber products into crumb is provided in Sections 6.2.2 to 6.2.4.

6.2.2 Cryogenic processes

This technique involves grinding waste rubber once it has been cooled to near its glass transition temperature and so is in the glassy state. In this state it can be ground using impact forces, and high-shear tearing forces are not required. The rubber is cooled by liquefied gases and, although air can be used, the most popular gas is nitrogen, which can either be applied to the rubber prior to grinding or during the process. The optimal grinding temperature has been reported as being just above the glass transition of the rubber [5]. Cooling of the rubber granules before grinding can be carried out in a number of ways, for example, directly on a conveyor belt, as it exits an extruder, in a tunnel-shaped chamber, or by immersion in a bath. Cooling during the grinding process is carried out by adding liquid nitrogen directly into the grinding mill. The types of mill that can be used are an impact mill or a hammer mill, and the crumb produced typically contains particles within the size range of 30–120 mesh that exhibit relatively smooth surfaces and, hence, lower surface areas than particles produced by ambient grinding operations. A photograph of rubber crumb produced using a cryogenic process is provided in Figure 6.1. This smoothness means that the untreated particles tend not to bind as well into other matrices (e.g., rubbers or thermoplastics) as ambiently ground crumb. The use of a cooling medium does provide benefits though: it reduces the temperature experienced by the rubber during the grinding process, and this means that less degradation occurs, particularly as the availability of oxygen is reduced. This removal of frictional heat also means that smaller particles can be produced then by ambient grinding, where there can be problems in getting down to mesh sizes of

Figure 6.1: Photograph of rubber crumb produced using a cryogenic process. Reproduced with permission from Smithers Rapra Technology Ltd, Shawbury, UK. ©2006, Smithers Rapra Technology Ltd.

100 and below, even with the use of partitioning agents to stop the particles sticking together (Section 6.2.4). Another particle property that varies between the two techniques is the width of the particle size distribution, with a broader distribution resulting from cryogenic grinding than ambient grinding. Cryogenic grinding can be carried out on rubber that has originated from vulcanised rubber, rubber products (e.g., tyres) and unvulcanised rubber.

An article has been published by Shell [6] that discusses the fundamentals of grinding rubber at low temperatures. It covers a wide range of subjects including the nature of the product produced, its required storage conditions, the economics of the process and the management and quality systems that need to be installed. The article also includes data on the particle size distributions of hydrogenated nitrile rubber (HNBR) crumb and the properties of fluorocarbon rubber (FKM) and HNBR compounds containing different amounts of rubber crumb.

6.2.3 Water processes

The crumbing of rubber can be achieved using water under very high pressure. Particles down to below 200 μm can be produced using this type of process. A photograph of rubber crumb that has been produced using a water jet process is shown in Figure 6.2. In common with cryogenic grinding, it has the advantage that the

Figure 6.2: Photograph of rubber crumb produced using a water jet process. Reproduced with permission from Smithers Rapra Technology Ltd, Shawbury, UK. ©2006, Smithers Rapra Technology Ltd.

cutting medium is capable of removing heat from the rubber, so reducing the possibility of degradation.

Two commercial processes that use high-pressure water jets to produce rubber crumb from waste rubber products are now reviewed. The Big Tyre Recycling Corporation (BTRC) situated in Gent, Belgium, has developed an ultra-high-pressure (UHP) water jetting system that is used commercially and has been described in a number of presentations and articles [7, 8]. By the use of UHP water jets, BTRC has shown that it is economic to remove the rubber in a single- step operation (referred to by BTRC as a crumbling process) from both the sidewalls and the tread of large off-the-road (OTR) tyres (e.g., open mining tyres with a diameter of approximately 4 m and weight of approximately 4 tonnes) and other large products, such as rubber crawler tracks for diggers or military tanks, heavy reinforced conveyor belts and so on (Figures 6.3 and 6.4). The capability to recycle large tyres is of particular value, because European Union (EU) Landfill Directive 1999/31/EC (European Community), which applies to waste tyres, does not encompass tyres that have diameters over 1.4 m. Due to a lack of economical recycling solutions, these tyres represent an environmental problem.

As an example, one type of jet used to strip the rubber from the tyres is 250 µm in diameter and the water pressure is 4,200 bar, with rotojets operating at 3,000 rpm. The process produces rubber crumb with an average dimension of 400 µm (~40 mesh), including fine dust particles as small as 1 µm. SEM micrographs were

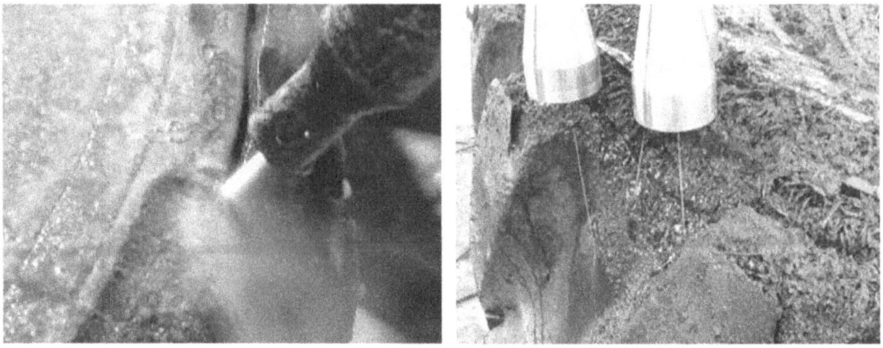

(a) (b)

Figure 6.3: The BTRC UHP water jetting process for removing rubber from end-of-life large tyres: (a) cutting operation and (b) jetting operation. Reproduced with permission of Big Tyre Recycling Corporation, Martelaarslaan, Belgium. ©2014, Big Tyre Recycling Corporation.

taken of the individual rubber grains and show them to have a complex morphology (Figure 6.5).

The recycled rubber crumb obtained by the BTRC UHP process can be vulcanised again, on its own or used in new compounds, providing that there is compensation for the sulfur present, in the same way as virgin NR or SBR. The BTRC process also removes the clean, high- value steel wire (steel needles used as concrete reinforcement) and the textile (fluff used as rubber reinforcement) (Figure 6.5).

Some of the other advantages claimed by BTRC for the process include:

- No limitation on the size of tyres
- Process requires no chemicals
- Crumb particles are activated by the process, which leads to matrix interactions
- 75% of the water used is recycled
- Very low energy required because the steel is not 'crushed'
- Relatively low investment and very low running costs

The BTRC recycled powder is used by companies to replace 40 phr of virgin NR in the tread of trailer tyres [8] and in all layers of solid tyres.

A UHP water jet system has been developed by a Hungarian company (HungaroJet Kft) and is the basis of a patented technology associated with the Hungarian Oil and Gas Company (MOL). In a paper presented by Kostyal [9] at the *Tire Technology 2014 Conference* in Cologne, the process was described as being the world's first continuous water jet milling plant. The process is stated as being capable of separating rubber from the tread, sidewall and inner layers of waste tyres to produce crumb that has a particle size below 400 µm with an extremely high surface-to-mass ratio and a combined convex/concave morphology that aids any interaction with a host matrix. In addition to tyres, the technology

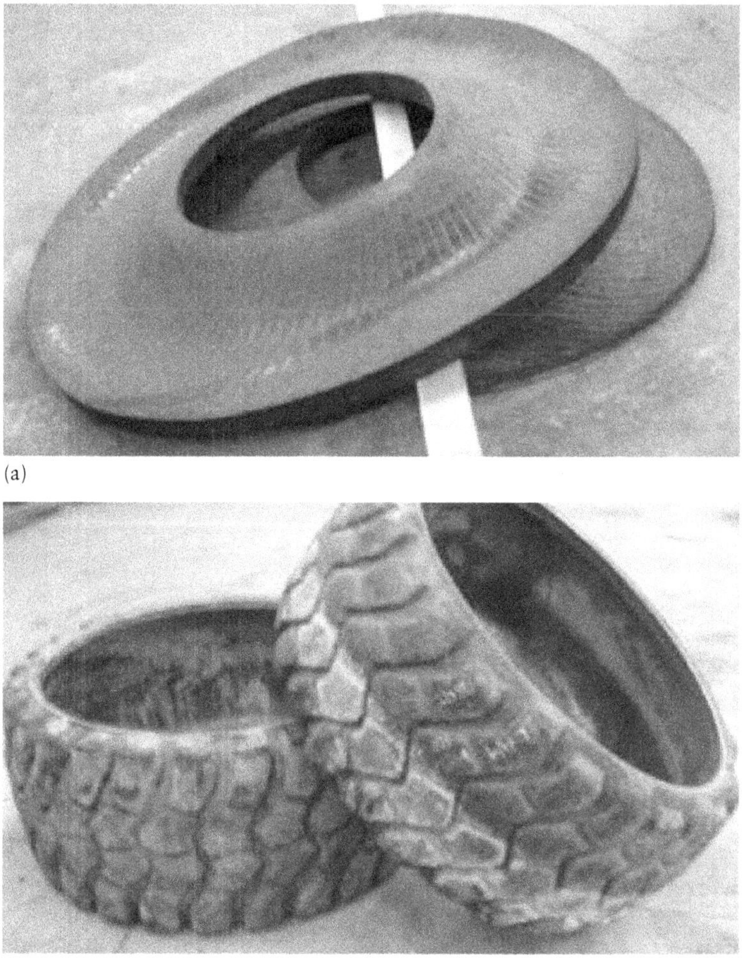

(a)

(b)

Figure 6.4: The BTRC UHP water jetting process: (a) side walls removed from large tyres and (b) tread sections removed from large tyres. Reproduced with permission of Big Tyre Recycling Corporation, Martelaarslaan, Belgium. ©2014, Big Tyre Recycling Corporation.

is capable of being used for other waste rubber products, such as those in the GRG sector.

Some of the other advantages that are claimed for this process are:

- Fine powder with a particle size of 50 μm is available.
- No surface oxidation of the crumb and no devulcanisation required prior to re-use.
- 100% metal and fibre-free product.
- Ability to selectively remove different parts of the tyres (i.e., tread, sidewall and butyl-based inner liner).

Figure 6.5: The BTRC UHP water jetting process: (a) recycled steel needles; (b) recycled fluff; (c) recycled powder, average 400 μm; and (d) SEM micrograph of recycled powder. Reproduced with permission of Big Tyre Recycling Corporation, Martelaarslaan, Belgium. ©2014, Big Tyre Recycling Corporation.

The HungaroJet process is being exploited commercially. A pilot plant with a capacity of 350 tonnes per year has been in operation since 2011 and is supplying leading European tyre manufacturers with crumb. The paper also describes how laboratory and industrial trials have shown that the crumb is capable of being used to produce high- quality thermoplastic rubbers, with polymers such as polypropylene, polyurethane rubber and polyvinyl chloride, and good quality rubberised bitumen materials for road-surfacing applications.

Another example of a type of process where water is used is disk grinding. The rubber is ground by friction created by it being passed between two grinding wheels. Water can be added to the process to both remove heat and improve its efficiency. The particle sizes generated by this process can be down to around 500 μm, without water, and down to 250 μm, or less, with water.

Water can also come into play to cool rubber that has been ground down to a fine size using air in a jet mill. The kinetic energy imparted to the granulate rubber that is fed into the jet mill is high enough that when the granules collide with each other or with parts of the interior of the mill, they disintegrate into smaller particles. There is a significant amount of heat build-up during the process, and cooling is necessary. As an alternative to the use of water as a cooling medium, liquid nitrogen can be used.

6.2.4 Ambient processes

Although the techniques in this section can be classified as methods that grind rubber under 'ambient' conditions, this is for convenience rather than a precise scientific definition, because any mechanical action on rubber will result in it heating up above ambient due to a combination of the inherent hysteresis in the material and its poor conduction of heat. The classification of 'ambient' is more to enable a distinction to be made between these types of processes and those where a medium (e.g., water or liquid nitrogen) is deliberately added to the process, or used in the process.

One of the types of mill that can be used for the ambient grinding of rubber is a cracker mill. In one type of design for these machines, one of the rollers has a corrugated surface and the rollers counter-rotate at a speed of around 50 rpm. There is a small gap between the two rollers, and this helps generate the high levels of friction required to grind the rubber. The material to be fed into a cracker mill is often pre-prepared in a shredder (Section 6.2.1), and the particle size generated by the mill is controlled by a number of factors, such as the size of gap between the rollers, the degree of friction between the rollers and the temperature of the rubber. Upon exiting the cracker mill, the ground rubber can pass through a gravity-fed 'cascade' of sieves in order to separate it into different particles sizes (e.g., 40 µm) prior to being bagged up and placed in a warehouse.

Rubber can also be ground at ambient temperatures by the use of any one of a number of extrusion-based processes. These can be twin-screw extruders that have been fitted with grinding elements and can be operated at conditions of high shear. Other extrusion processes can deliberately heat up the rubber, so that it is subjected to both high shear and high temperatures. These types of processes can be regarded as having a dual role, as they can also start to devulcanise the rubber, particularly those that are sulfur-cured, as the sulfur–sulfur crosslinks are both more thermally labile than main-chain carbon–carbon bonds and are less flexible, and so will preferentially break under conditions of shear (Chapter 4, Section 4.2). The occasions when these types of extrusion processes are used more for devulcanisation than particle size deduction are discussed in Chapter 4, Section 4.4.

If the starting material is a composite one, such as tyres, it has been found that ambient grinding processes tend not to remove as much fabric and metal from the final crumb product as cryogenic processes. For example, an ambient ground rubber has been found to have a fibre content of 0.5% and a steel content of 0.1%, whereas that produced by a cryogenic process had levels below the detection limit for these two components [10, 11].

The rubber particles produced by ambient-type grinding processes tend to have a narrower particle size distribution than, say, cryogenically ground rubber, and have a more complex morphology and hence larger surface area. A photograph of rubber crumb that has been produced using an ambient process is shown in Figure 6.6. This relatively complex surface morphology and larger surface area has, for a number of years, been regarded as being responsible for the better bonding characteristics that are achieved with ambient ground crumb when it is mixed with binders and other materials to form composites. A short paper by Winters in 1986 [12] commented on the difference between ambient and cryogenically ground rubber in this respect, while also discussing some of the products (e.g., sports surfaces, asphalt-based surfaces, and extruded products) that can be constructed using ground rubber particles.

Figure 6.6: Photograph of rubber crumb ground using an ambient process. Reproduced with permission from Smithers Rapra Technology Ltd, Shawbury, UK. ©2006, Smithers Rapra Technology Ltd.

Ambiently ground rubber particles can also have a more complex surface chemistry due to some oxidative degradation having taken place at their surface. In general, it is also harder to generate the smaller particle sizes (e.g., <100 mesh) using ambient-type processes because the increasing amounts of heat generated as the particle size is reduced can cause too much degradation and the particles stick together, even if a partitioning agent is used. Because of this, one of the factors that determines what size a particular rubber can be ground down to is its thermal stability. For example, NR and SBR, which have relatively poor thermal stabilities, can be difficult to grind below 80 mesh, whereas FKM, which has very good thermal stability, can be ground down to at least 120 mesh. Another factor that can come into play is the modulus (or hardness) of the rubber, with the lower-modulus rubbers (e.g., NR and SBR) deforming more under the action of the grinding mechanism, and so generating more heat, than the higher-modulus rubbers, such as the fluorocarbons.

With respect to the use of partitioning agents to stop rubber crumb particles sticking together, care has to be taken in their use, as too high a loading can lead to them interfering with subsequent devulcanisation processes and uses for the crumb that require it to have good interfacial bonding into a host matrix (e.g., plastic) or with itself (e.g., cryosintering processes).

Fine rubber particles can be produced from waste rubber at ambient temperature using a process called solid-state shear extrusion (SSSE). Chaubey and Arastoopour [13] have described how they have used an extended screw within an SSSE machine to achieve a greater degree of pulverisation of the rubber, a higher throughput, and reduced power consumption. They found that controlling the residence time of the particles in the pulverisation section of the machine was important to produce fine particles that did not agglomerate.

A company called Innovative Polymer Technologies has developed a solid state shear pulverisation process for powder production, intimate mixing and compatibilisation of polymer blends, including waste rubber–thermoplastic blends [14]. The process creates powders with a large surface area and complicated morphologies. In addition to waste rubber, the process is also capable of producing powders from thermoplastic rubbers, and waste rubber–thermoplastic blends. The resulting powders can be used for a variety of applications.

A novel variant on ambient grinding technologies was described by Asadauskas and co-workers [15] at the *Tire Technology 2014* event in Cologne. In this high-shear process, both mechanochemically effective and tribologically effective additives are employed. These additives are claimed to provide the following twin benefits:

- 95% conversion of tread buffings into 20 mesh powder (tribological additives).
- 90%+ reduction of crosslink density by devulcanisation (mechanochemical additives).

The properties of the fine, devulcanised rubber crumb that results from the process varies according to the types of additives used and their levels, and the crumb has been shown to be used effectively to replace virgin rubber in new rubber compounds. For example, experimental trials have been undertaken with rubber crumb obtained using this technology from car tyre treads incorporated into SBR compounds. The results that were obtained on the resulting vulcanisates were compared to those produced using conventional rubber of the same mesh sizes, and they demonstrated major improvements in tensile strength and other key properties.

6.3 Sieving of rubber crumb

Once the rubber has been granulated, the next step is usually to sieve it into products that have well-defined particle sizes and particle size distributions. This operation confers the dual benefit of enabling a producer to deliver consistent high-quality material to a customer who has developed and optimised a process (e.g., a devulcanisation process), and enabling the provision of different grades of material, helping to ensure that the potential number of end-uses is not restricted.

The sieving stage can be carried out using multiple sieving stations immediately, or very shortly, after the granulation stage to maximise the efficiency of the overall process. The particle size distribution of the material that is produced by the sieving operation can be characterised using a standard such as ASTM D-5644–2008 (determination of particle size distribution; Section 6.4.2.2).

Sieved material will normally be sold by specifying a property such as the amount of material (in weight percent) that is retained on a particular set of sieves. For example, a rubber crumb product that had been produced from truck tyre treads could be sold as a '40 mesh material', in which nothing had been retained on a 25 mesh sieve (0.60 mm) and no more than 3% had been retained on a 35 mesh sieve (0.425 mm). In other words, at least 97% of the product had a particle size of less than 35 mesh (0.425 mm).

6.4 Quality control monitoring of rubber crumb and products containing rubber crumb

Rubber crumb is an important resource, but as described above can be derived from a number of different waste rubber sources and can vary in quality, particularly the amount of contamination present. So, in order to ensure that the market for this material is not hindered by lack of confidence in the consistency of its quality, a number of standards and protocols have been published in both Europe and the USA. Some examples of these are shown below. The Protocol published by WRAP (Section 6.4.1) addressed rubber crumb from tyres, but also had a wider scope and includes other types of rubber products from waste tyres (e.g., rubber shred). In addition to the inherent quality and properties of the rubber crumb, consideration also has to be given to its storage, and the quality of the final products, particularly engineering products, that it has been used to manufacture.

The use of rubber crumb in polymer products, for example in blends with other rubbers, thermoplastics or thermosets, will tend to be assessed and controlled by the test specifications that those particular products have to meet with regard to important physical, chemical and long-term durability properties.

6.4.1 The WRAP quality protocol and guidance documents

6.4.1.1 UK WRAP quality protocol

In November 2009, WRAP published a Quality Protocol for tyre- derived rubber materials. This Quality Protocol was produced using funding provided by DEFRA and the Welsh Assembly Government (WAG) as a business resource efficiency activity. It was developed by the Environmental Agency and WRAP in consultation with DEFRA, WAG, industry and other regulatory stakeholders. The Quality Protocol is only applicable to England and Wales in the UK, and sets out end of waste criteria for the production and end-use of tyre- derived rubber materials. The main stages and control mechanisms of the Quality Protocol are shown in Figure 6.7.

The reason why the Quality Protocol has been published is related the EU Waste Framework Directive (2006/12/EC) and the importance of defining at what point rubber waste has been fully recovered and ceases to be waste as it is defined in Article 1(1)a of this Directive. Hence, this is also the point at which the Directive no longer applies and so it can be used without the need for waste management controls. It was perceived that problems in defining this point had inhibited the development and marketing of materials produced from waste rubber.

The main purposes of the Quality Protocol are:

- To clarify the point at which tyre-derived rubber materials cease to be waste and so waste management controls no longer apply;
- To provide users with confidence that the tyre-derived material conforms to an approved material standard;
- To provide users with confidence that the tyre-derived materials are suitable for use in designated applications including by conforming to an engineering standard;
- To protect human health and the environment; and.
- To describe an acceptable good practice for the use of tyre-derived rubber materials.

There can often be some debate regarding the definition of rubber granules, rubber powder and so on, and the protocol provides some assistance here by providing the definitions shown in Table 6.1.

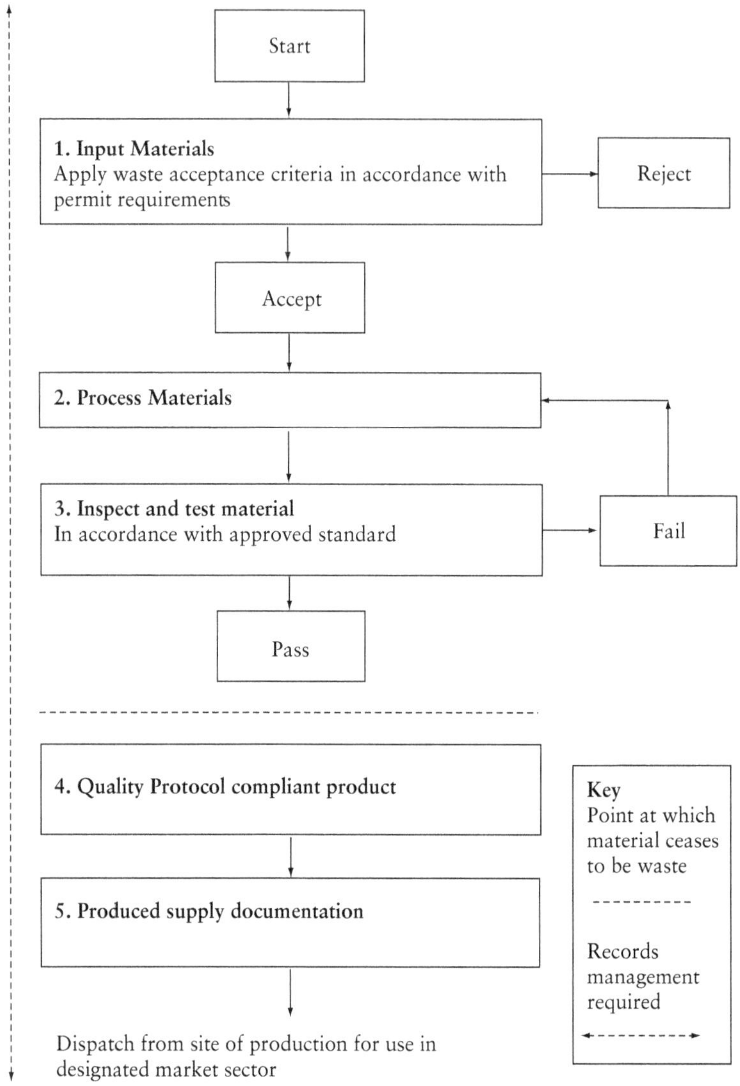

Figure 6.7: Main stages and control mechanisms of the Quality Protocol. Reproduced with permission from the WRAP Quality Protocol, Waste and Resources Action Programme, Banbury, UK, p.7. ©WRAP.

Note: WRAP does not warrant that the functions contained in the material contained in this Site will be uninterrupted or error free, that defects will be corrected, or that this Site or the server that makes it available are free of viruses or bugs or represents the full functionality, accuracy, and reliability of the materials.

To comply with the Quality Protocol the tyre-derived rubber materials must be destined for use in one if the designated applications within certain designated market sectors:

- Civil engineering (unbound applications):
  - As a replacement aggregate in the construction of road infrastructure, e.g., as roadbed stabiliser, slope stabiliser and as an additive in asphalt.
  - As a loose or bound material in surfacing, e.g., of footpaths, cycle paths and railways as ballast.
- Civil engineering (bound applications):
  - As a replacement aggregate in the construction and building industry, e.g., in block fabrication.
  - As a bound rubberised product, e.g., wall and floor boarding, street furniture (e.g., seating) and railway crossing surface matting.
- Landfill engineering:
  - As a replacement aggregate in the construction of landfill sites, e.g., as a drainage layer.
- Sports, recreation and leisure applications:
  - As unbound material in sports surfacing, e.g., artificial turf, racecourses and running tracks.
  - An unbound material in recreation and safety surfacing, e.g., playground surfaces, nature trails, bridle ways, cycle trails.
  - As bound material in safety surface matting, anti-vibration matting, impact protection barriers.
- Industrial and consumer applications (bound applications):
  - Automotive, e.g., new and remoulded tyre, and moulded parts.
  - Street furniture and road surface matting (e.g., level crossings).
  - Rubber matting, e.g., for livestock.
  - Ingredients in building products, e.g., rubberised roof tiles, rubberised adhesives and mastics.
  - Consumer products, e.g., footwear.
  - Carpet underlay.

For the civil engineering and landfill engineering market sectors, there may be applicable standards in the relevant categories listed in Section 6.4.3.2.

The Quality Protocol also provides information in a number of other areas that are relevant to tyre-derived rubber materials, such as the good practice guidelines shown in Sections 6.4.1.3 to 6.4.1.5, and these are present in Appendices in the Protocol.

6.4.1.2 Good practice for the storage of rubber crumb

Good practices for the storage of rubber crumb and tyre-derived materials should be followed, both on site and at an interim location, and these can be found in Annex B of PAS 107:2007 (Publically Available Specification prepared by the BSI).

Additional good practice guidance was provided in the WRAP Quality Protocol described above.

A summary of Annex B of PAS 107:2007 can be provided as follows:

- Processors should retain ownership of all tyre-derived materials stored at any location until such time as the materials are delivered to the customer at the site of their final use.
- Tyre-derived rubber materials should only be stored at an interim location in maximum amounts of 5,000 tonnes for a maximum of 12 months.
- Tyre-derived rubber materials may be at risk from fire and spontaneous heating and fire safety guidance from the UK Health and Safety Executive should be followed and measures taken to mitigate the risk of spontaneous heating, e.g., minimising pile size and protecting from direct sunshine.

6.4.1.3 Good practice for unbound applications of tyre- derived rubber materials in sensitive environments – UK WRAP quality protocol

- The limit of working within 10 m of a main river watercourse set out in the Pollution Prevention Guideline PPG5 published by the Environmental Agency should be followed.
- Tyre-derived rubber materials should not be used in areas with high pH (e.g., pH 8 or above) or very low pH (e.g., pH 5 or below) as there is a greater potential for metal/organic mobilisation.
- Tyre-derived rubber materials should not be used for high load- bearing applications as their ability to leach chemicals increases.

6.4.1.4 Good practice for unbound, sports, recreation and leisure applications in close proximity to aquatic receptors – UK WRAP quality protocol

- The limit of working within 10 m of a main river watercourse set out in the Pollution Prevention Guideline PPG5 published by the Environmental Agency should be followed.
- Steps should be taken to ensure that no pathway exists for surface water run-off directly into water courses. If this is not possible and/or a potential pathway is found to exist, steps that should be taken include, for example:
  - Containing any leachate and treating in the same manner as sewage.
  - Where practical containing all loose tyre-derived rubber materials using appropriate barrier methods (e.g., kerbing) to prevent any free release into the environment.
- Records should be kept and retained for a minimum of four years. The guidance notes provide details of what needs to be recorded, for example:
  - The location of sites.
  - The dates of spreading.

6.4.1.5 Good practice for unbound civil engineering applications of tyre-derived rubber materials – UK WRAP quality protocol

- The limit of working within 10 m of a main river watercourse set out in the Pollution Prevention Guideline PPG5 published by the Environmental Agency should be followed.
- Steps should be taken to ensure that no pathway exists for surface water run-off directly into watercourses. If this is not possible and/or a potential pathway is found to exist, steps that should be taken include, for example:
  - Developing suitable containment mechanisms and/or leachate collection systems.
  - Ensuring adequate dilution at the receptor, e.g., fast-flowing watercourse.

6.4.2 Quality standards

6.4.2.1 The production of rubber crumb

With regard to the manufacture of crumb using ambient grinding- type processes, the BSI has published a PAS for the collection, initial storage, production and final storage of size-reduced, tyre-derived rubber materials. It is designated PAS 107:2007. This document has already been referred to in Section 6.4.1 in connection with quality protocols that apply to rubber crumb.

The use of cryogenic grinding and water jet processes for the manufacture of crumb is not currently included in any PAS, but PAS 107:2007 can be used for these types of processes when manufacturers want their crumb to be considered as having been manufactured according to the Quality Protocol published by WRAP (see Section 6.4.1). The only part of PAS 107:2007 that does not need to be adhered to is Section 7.3 (relating to the production process), which only applies to ambient grinding processes.

6.4.2.2 Properties of rubber crumb

In Europe, the Comité Européen de Normalisation (CEN) technical committee TC 366 (Project Committee – Tyre recycling) published a Technical Specification (TS), CEN/TS 14,243:2010 in 2010 that provides a definition for the categories of materials that are produced from end-of-life tyres based on their dimensions or impurities. This publication also provides test methods for the determination of the dimensions of the materials produced from all categories of end-of- life tyres, at all stages within the treatment process, as well as for the determination of impurities. It also describes test methods for the collection of samples, and for the preparation of representative samples based on a sampling plan for the purpose of determining dimensions and impurities. It does not cover the following technical areas:

- Operational performance or the fitness of the materials for the desired end-use, which has to be agreed between the producer and the customer.
- Downstream products prepared from the materials, including reclaimed or devulcanised rubber, surface-modified powders or pyrolysis products.

American standards that address the general properties of rubber crumb include ASTM D-5603 (Classification for Rubber Compounding Materials – Recycled Vulcanisate Particulate Rubber). This standard has sections covering the classification of rubber crumb into six classes – (1) to (6) – covering the determination of mesh size, the determination of bulk density and other key properties.

As described in the sections above, once a grinding process (ambient, cryogenic or water jet) has produced crumb it is sieved, which produces a nominal size, e.g., 40 mesh (equivalent to ~400 μm).

However, another important property of the crumb will be its particle size distribution. This can be determined using ASTM D-5644 (Standard Test Methods for Rubber Compounding Materials – Determination of Particle Size Distribution of Recycled Vulcanisate Particulate Rubber). Two different methods are described in this standard, Methods A and B:

- *Method A*: This describes the Ro-tap sieve test method for rubber particles that are 60 mesh or coarser. This method is based on the size designation screen, and sets screen retention limits of a maximum of 5% for up to 850 μm (20 mesh) particles, a maximum of 10% for 600–150 μm (30–100 mesh), and a maximum of 15% for 128–75 μm (120–200 mesh). No rubber particles are allowed to be retained on the zero screen.
- *Method B*: This method differs from Method A in that it is the number of particles of a particular size that is determined and not the weight percent retained on a specific screen. The size of the particles is assessed using optical microscopy and an experimental approach that is based on Test Method ASTM D-3849 (Test Method for Carbon Black – Morphological Characterisation of Carbon Black using Electron Microscopy).

Method B also differs from Method A in that it is an ultrasonic technique and the use of this technology addresses the problems that may be caused by tackiness and the static electrical forces that recycled rubber particles may exert on each other to form agglomerates. The problem of agglomeration can be particularly prevalent for particles that are 80 mesh or finer. The use of ultrasonic agitation eliminates the formation of agglomerate by dispersing the particles.

Standards such as ASTM D-5603 and D-5644 deal with the physical properties of crumb, but there are also the chemical properties to consider, and one standard that addresses these is ASTM D-297. This includes experimental methods to determine a host of chemical properties, such as the sulfur content of the crumb, the ash content (i.e., total inorganic content), and the chemical composition of the ash, providing a guide to the types of inorganic fillers present in the crumb.

In addition to the traditional methods set out in the ASTM standards, it is also possible to use instruments that are generically known as particle analysers for the characterisation of rubber crumb. These instruments can characterise important properties such as particle size, particle distribution and particle morphology. They can also enable individual particles from each sample to be fully characterised by measuring a number of morphological parameters and their associated distributions. These parameters include particle diameter, width, length, aspect ratio, circularity, convexity and elongation, and together they provided a 'fingerprint' for each sample. In addition to these morphological properties, such an instrument can also provide information on the 'volume' distributions for these key parameters: D0.1, D0.5 and D0.9. The values for these three parameters are the size values corresponding to the cumulative particle distribution at 10, 50 and 90%. Thus, the D0.1 value represents a size value below which 10% of the cumulative distribution is present, and D0.5 and D0.9 represent size values below which 50 and 90% of the cumulative distribution occurs. From these definitions, it can be seen that the D0.5 value can be regarded as the median size value.

Some examples of the data that can be provided by these types of instruments are shown in Tables 6.2 to 6.4. In these tables, samples of rubber crumb were produced by three different grinding techniques (ambient, cryogenic and water jet) and were analysed using a Pharma Vision 830 Advanced Image analyser by Smithers Rapra. Representative examples of images captured for these three samples with this instrument are shown in Figures 6.8 to 6.10.

Table 6.2: Rubber crumb produced by an ambient grinding process.

Parameter	Volume distribution (µm)			Mean (µm)
	D0.1	D0.5	D0.9	
Elongation	564	1103	1602	258
Roundness	0.26	0.60	0.82	0.51
Convexity	0.81	0.89	0.94	0.91

Note on observations made during analysis: the particles were found to have many different shapes, many of which were irregular and inconsistent in appearance. Most of the particles had a jagged or irregular surface morphology (Figure 6.8)

6.4.2.3 The use of rubber crumb in civil engineering and landfill engineering applications

If rubber crumb is going to be used in these types of applications, the producers of the crumb should be able to demonstrate compliance with the appropriate engineering standards to ensure that the material is suitable for the intended application. Some examples of the most relevant standards are shown below.

Table 6.3: Rubber crumb produced by a cryogenic grinding process.

Parameter	Volume distribution (µm)			Mean (µm)
	D0.1	D0.5	D0.9	
Elongation	409	656	990	432
Roundness	0.25	0.59	0.84	0.54
Convexity	0.86	0.93	0.97	0.92

Note on observations made during analysis: the particles of rubber were found to be generally very cuboid in appearance and many appeared to have well-defined straight shaped edges, corners and a smooth surface. A small number of rod-like particles were also evident in the sample (Figure 6.9)

Table 6.4: Rubber crumb produced by a water jet grinding process.

Parameter	Volume distribution (µm)			Mean (µm)
	D0.1	D0.5	D0.9	
Elongation	197	510	1,249	181
Roundness	0.05	0.39	0.76	0.58
Convexity	0.79	0.87	0.94	0.94

Note on observations made during analysis: the particles were found to be very uniform in size and fairly consistent in their particle shape. The particles tended to have more of an irregular surface morphology when compared to the ambient and cryogenic crumb samples, and there were relatively few particles that differed widely in shape (Figure 6.10)

Civil engineering applications of crumb:

- British Standard European Norm (BS EN) 13,043:2002: aggregates for bituminous mixtures and surface treatments for roads, airfields and other trafficked areas.
- BS EN 13,055–2:2004: Lightweight aggregates part 2: lightweight aggregates for bituminous mixtures and surface treatments for unbound and bound applications materials.
- BS EN 13,242:2002: aggregates for unbound and hydraulically bound materials for use in civil engineering work and road construction.
- BS EN 13,285:2003: unbound mixtures: specifications.
- BS EN 13,450:2002: aggregates for railway ballast.

Landfill engineering applications of crumb:

- Guidance on the use of tyres for leachate drainage blankets on landfill sites (Environment Agency 2008).

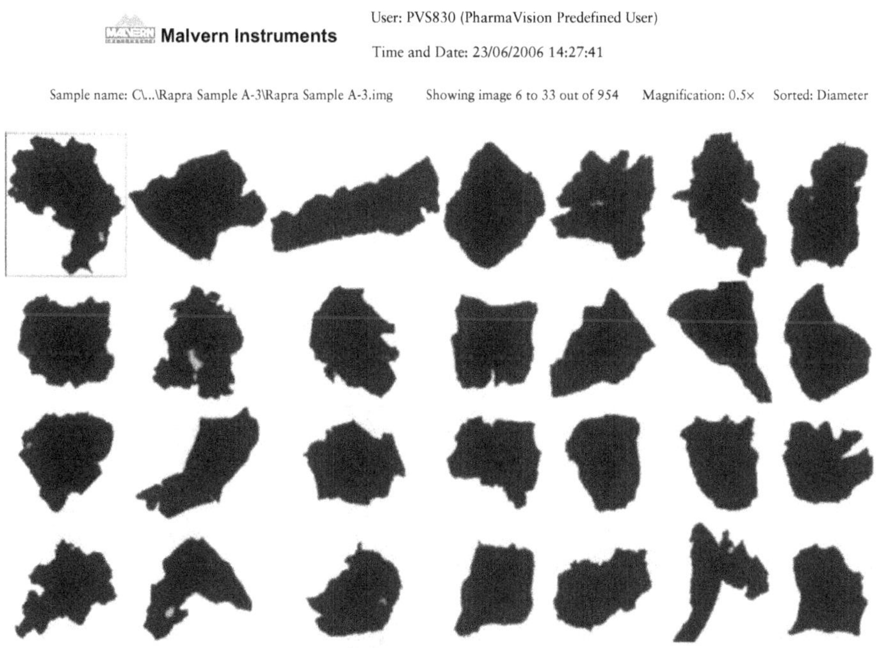

Figure 6.8: Profile images generated using a particle analyser of rubber crumb produced using an ambient process. Reproduced with permission from Smithers Rapra Technology Ltd, Shawbury, UK. ©2006, Smithers Rapra Technology Ltd.

6.5 Sintering techniques to produce products from rubber crumb

The use of high-pressure sintering processes to consolidate rubber crumb (often waste tyre crumb) into new rubber products has been evaluated by a number of workers and has been carried out for many years. In common with a number of the devulcanisation processes described in Chapter 4, these investigations have built on the work carried out by Tobolsky and co-workers [16] and Murakami and Ono [17]. Tobolsky realised that sulfur–sulfur bonds in rubbers are relatively weak and are responsible for a lot of the degradation that takes place. He used elevated temperatures to study the rearrangement of these bonds in a number of rubber compounds to determine the rupture and reformation behavior, and found that the rates of bond scission and reformation are essentially equal when the materials are protected from oxygen and other reactive compounds. When a sintering process is used to create a moulded rubber article, the pressure in the mould forces the interfaces of the rubber particles together and the chemical rearrangement of the sulfur–sulfur bonds in the rubber welds them together.

This approach to re-using waste rubber has been used for over 30 years. For example, in the early 1980s research workers in France, Acetta and Vergnaud [18],

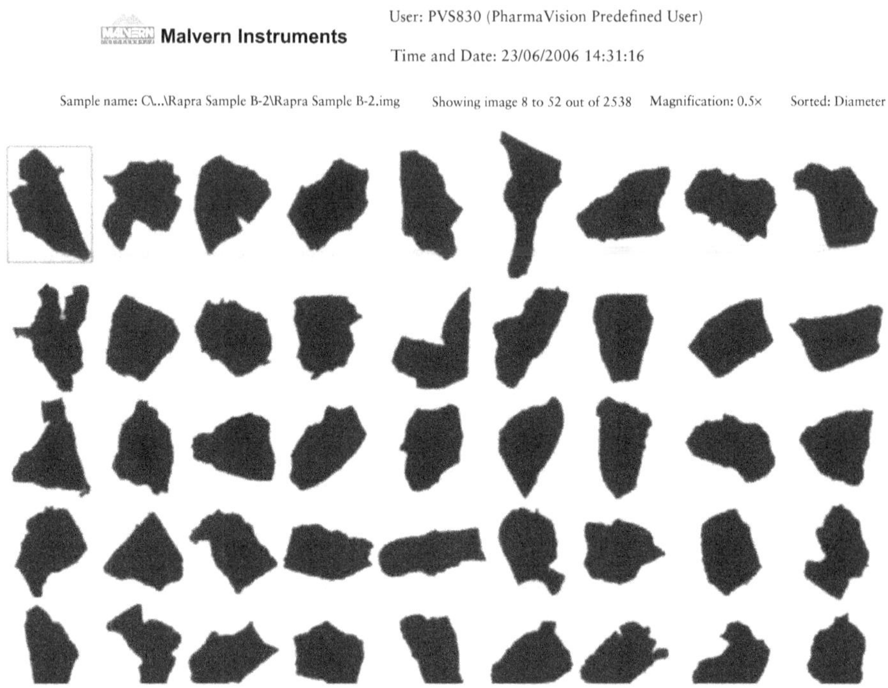

Figure 6.9: Profile images generated using a particle analyser of rubber crumb produced using a cryogenic process. Reproduced with permission from Smithers Rapra Technology Ltd, Shawbury, UK. ©2006, Smithers Rapra Technology Ltd.

carried out a series of experiments that involved taking cryogenically ground tyre rubber (300–900 μm particle size range), mixing it with readily available rubber chemicals (e.g., sulfur and dioctylphthalate plasticiser) and then pressing the mixture together in a heated mould. They reported that the best conditions were found to be 10 min heating at 180 °C using a moulding pressure of 10 MPa. One of their main conclusions was that, for the production of vibration damping devices and products, the amount of sulfur in the starting mixture had an important influence on the level of vibration absorption achieved. The levels of tensile strength and the hardness that were obtained were also dependent on the level of sulfur. They considered that the vibration damping products could have a use in industry to absorb the vibrations that are generated by machinery, and that their ability to absorb impact noise could be utilised in flooring products, possibly in constructions where rubber pads made from sintered ground rubber were overlaid with mortar [19].

One of the attractions of sintering processes for re-using waste rubber is that they can be relatively simple and cost-effective, as they require only minimal preparation of the crumb and can be carried out using conventional compression

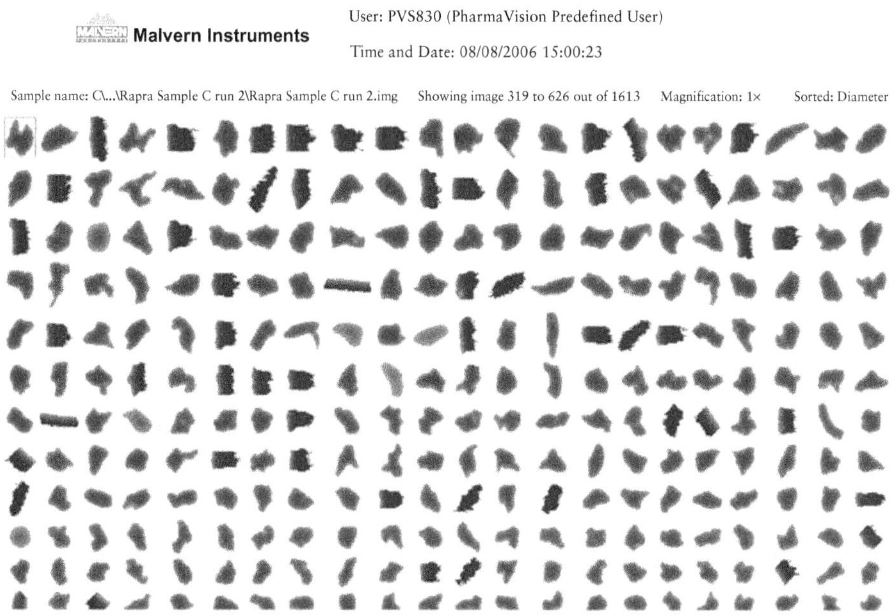

Figure 6.10: Profile images generated using a particle analyser of rubber crumb produced using a water jet process. Reproduced with permission from Smithers Rapra Technology Ltd, Shawbury, UK. ©2006, Smithers Rapra Technology Ltd.

moulding techniques. The drawbacks are that it is usually a relatively slow batch process, and, unless a high level of quality control is employed in areas such as the moulding conditions and the degree of contamination in the rubber crumb, poor-quality products can result.

Where a cryogenic process is initially used to generate crumb, this technique can be referred to as 'cryosintering'. However, this terminology is not always exclusively applied to the sintering of crumb that has been produced by a cryogenic grinding process and, rather confusingly, it can be used for the sintering of crumb that has been produced by any of the possible routes, i.e., ambient, water jet or cryogenic. An instance of this occurring was in the case of a EU-funded Framework 6 project called 'Cryosinter' that ran from 2006 to 2008. This project evaluated the sintering of rubber crumb produced using the three main generic crumbing techniques mentioned above. The project looked at how a number of variables affected the final properties of the sintered products, including:

- Particle size and morphology of the rubber crumb
- Type of rubber crumb, e.g., crumb from truck tyre tread or whole tyre crumb
- Sintering conditions, e.g., temperature and time
- Influence of different types and mixtures of crumb activating chemicals

One of the primary goals of the project was to see if the technique could be used to produce materials that were of a sufficiently high quality to be used to manufacture commercial items such as shoe soles. Overall, the project made good progress towards its goals, and some of the salient points that were established included:

- A tensile strength above 10 MPa and an elongation at break above 250% are possible in the final sintered products when chemical agents are added to improve the interfacial interaction between the crumb particles.
- The quality of the crumb with regard to contamination such as metal and fabric (from the tyre carcass) is very important, as small amounts of these can adversely affect important final properties such as tensile strength and elongation.
- The particle size distribution and surface morphology are important. In general, a wider particle size distribution and a coarser particle surface favour good-quality sintered mouldings.
- The presence of a partitioning agent (e.g., a stearate) on the surface of the rubber particles can interfere with the interfacial bonding of the particles, resulting in a loss of mechanical properties.
- The process conditions are important, with improvements in final properties sometimes being achieved by using a shorter sintering time to avoid degradation of the material in general, and the interfacial areas in particular.
- The use of injection moulding, as opposed to the more conventional approach of compression moulding, showed some promise. If this could be achieved commercially it would greatly improve the productivity of the process.

As the Cryosinter project members and other researchers have found, adding chemical agents to the rubber crumb prior to the sintering process is usually necessary in order to achieve the best results in the final product. A number of different types of chemical agent have been evaluated for the sintering of rubber particles:

- Difunctional 'bridging compounds', e.g., phthalimide, *N*-methyl phthalimide and maleic anhydride.
- Organic acids, e.g., adipic acid, benzoic acid, phthalic acid, maleic acid and salicylic acid.
- Conventional rubber chemicals such as sulfur and mixtures of sulfur and rubber curing accelerators (e.g., tetramethyl thiuram disulfide, TMTD).
- Peroxides, e.g., dicumyl peroxide.

A lot of work in this field has been carried out by a team in Massachusetts in the USA, who have published a number of papers and articles [20–23]. Their work includes a wide range of different types of rubber crumb, from the diene types (e.g., NR, SBR and EPDM) to more specialist materials not vulcanised using sulfur-based cure systems, such as FKM, polysulfide rubber and silicone rubber. Their programme has also encompassed a large range of chemical agents and other variables, including the moulding conditions and particulate properties of the rubber powders (e.g., different particle size distributions). They concluded from the results of their investigations that the high-temperature, high-pressure sintering route was an ideal method for making value-added products (e.g., gaskets, O rings, low-end tyres and roofing materials) from waste rubber and that the properties of the moulded products were capable of comparing well against some typical virgin rubber compounds. An understanding of the mechanism of consolidation/ particle adhesion was also obtained, with the most likely explanation involving bonds breaking into radicals with an increase in mobility allowing the polymer chains with radicals to cross the particle interface and react with other radicals to form new bonds. They considered that this process was also happening within the particles, but that this did not affect the mechanical properties or particle adhesion. Sulfur was also found by the group to be capable of accelerating this process.

An example of research where sulfur, an accelerator (TMTD) and other ingredients (some of an undisclosed nature) have been used to prepare rubber blocks from sintered waste tyre crumb, for end-uses that could include ballast mats for high-speed trains, is that carried out by Kim and Lee [24]. They cured the blocks using a 50 tonne hydraulic press for 15 min at 150 °C and one of their aims was to compare the physical properties of their materials with sheet products that had been produced from tyre crumb using polyurethane and SBR binders. Their results showed that their materials had poorer tensile strength, elongation at break and ageing performance than the other two products, but the cost of the blocks could be cut by up to 50% by not using a binder.

A team of workers in the USA have explored the possibilities provided by a two-stage recycling process to re-use rubber waste generated by a commercial rubber company [25, 26]. The rubber waste had the following composition:

- Standard Malaysian Rubber-20 (NR): 53.9%.
- Semi-reinforcing furnace carbon black: 26.9%.
- Aromatic oil: 10.8%.
- Vulcanising systems and other additives: 8.4%.

The first stage was an SSSE process (Section 6.2.4) that pulverised the rubber waste into small particles, with ~60 wt% having a size of less than 600 μm. A moulding stage (i.e., the sintering stage) was then used to press the rubber powder into sheets in the absence of any virgin rubber binder. A large range of experiments were carried out by the workers to investigate the effect of varying a number of process parameters on the properties of the sintered product, including:

- Extracting the powder with a solvent (toluene) prior to moulding.
- Use of different moulding temperatures, times and pressures.
- Addition of sulfur in isolation, or with other chemicals, e.g., *tert*-butyl-2-benzothiazole sulfenamide and a pre-vulcanisation inhibitor.

In addition to characterising the rubber powder and the sintered products using physical and chemical techniques, they also used an SEM to analyse the surfaces of sintered and control (i.e., cured virgin material) products before and after they were fractured in a tensometer. Their results showed that the properties of the sintered products were inferior to those of the virgin material due to poor particle bonding, but that they were sufficient for the production of items such as mats, pads, carpet underlay and so on. They found that the moulding conditions and the properties of the powder (e.g., particle size distribution) had a significant effect on the properties and the crosslink densities that could be obtained. Higher moulding temperatures and pressures generally improved the interfacial bonding of the rubber particles, and a high proportion of large, coarse particles with a high crosslink density within the particles, inhibited it. The possibility of using the rubber powder produced by the SSSE process at low levels of addition in the manufacture of tyres was also postulated by the workers.

Workers in Russia [27] have carried out sintering experiments using rubber powder that had been prepared by high-temperature shear deformation of EPDM vulcanisates with different crosslink densities. They prepared and tested samples that had been sintered under different conditions. They also blended the EPDM powder with virgin EPDM and tested the blends. The performance of the EPDM powder was found to be dependent on particle size, surface area and composition.

Work carried out in Poland [28] has shown how treating comminuted scrap rubber with standard rubber industry ingredients, such as sulfur and rubber accelerators, and subjecting the resulting material to high temperatures and pressures can create a range of useful industrial and consumer products in a cost-effective way.

6.6 Activation of rubber crumb

Increasing the activity of the surface of rubber crumb can yield big improvements in the properties of the products that are produced when it is mixed and blended into virgin rubber, thermoplastics, thermosetting resins or with other materials (e.g., cement).

Some of the other recycling methods that have already been reviewed in this book can be regarded as involving, to varying degrees, the surface-activation of rubber crumb, for example the approaches grouped under the term 'sintering techniques' in Section 6.5. The microbiological devulcanisation processes (Chapter 4, Section 4.8) could also be regarded as another example, as the devulcanisation that takes place tends to be concentrated at the surface of the rubber particles. Different methods and approaches for the activation of rubber crumb are also encountered in the section of this book that reviews the many different products in which this material can be used (Chapter 7). The surface activation that occurs in these cases is usually integral and specific to the re-use technology concerned, whereas this section briefly mentions some of the more general surface-activation processes, or includes specific activation technologies that have been the subject of scientific papers and articles.

There are a number of general techniques that can be used to enhance the surface activity of rubber crumb, including:
- Compatibilisation or activating agents, e.g., carboxylated polymers, amines and so on.
- Grafting polymers onto the surface, e.g., ethyl acrylate rubber.
- Polymeric binders, e.g., polyurethanes and acrylates.
- Chemical treatments to raise the surface free energy, e.g., chlorination.
- Plasma processes, e.g., using oxygen, ammonia or hydrogen atmospheres.
- Corona discharge and electron-beam processes.
- Irradiation with microwaves.
- Extraction with hot water (~90 °C) to remove oils and chemicals from the surface.
- Mechanical activation on a mill at relatively low temperatures (e.g., <100 °C).

Some of the specific techniques that have been published in the literature are described below, for example, a technique that has been developed by Vredestein Rubber Recycling involves the mixing of rubber crumb with rubber-curing additives and a crosslinkable polymer. This process, called the Surface Activated Rubber Crumb (Surcrum) process, is capable of producing final products that have better physical properties than if the crumb is only mixed with the curing additives. Using this technology, the company developed high-molecular-weight polymer dispersions of 0.4 mm rubber crumb derived from a number of rubbers, for example NR, SBR, nitrile rubber (NBR) and chloroprene rubber (CR) [29, 30].

Stark and co-workers of the US company Rubber Research Elastomerics reported on the use of polymers to activate rubber powder in the mid-1990s [31]. In order to interact effectively with the crumb, the polymers need to have a low viscosity at the processing temperatures that are being used and this requires that they have a molecular weight in the range of 1,000–100,000 g/mol. The resulting polymer dispersion was usually dried before being added to the rubber crumb, and surface activation systems using this technology have been developed for rubber crumb with a particle size of 0.6 mm that has been derived from a number of sources. Examples of the types of crumb include whole tyre, truck tyre tread, EPDM, NBR, isobutylene-isoprene rubber and CR, and the activated crumb can be used in both new rubber compounds and in other materials.

A novel 'surface polymerisation' process has been evaluated in a research project for WRAP by a Barnsley-based company called Rubber Recovery and

Envirogen Technologies Ltd, which is the global development partner for Rubber Research Elastomers Inc. The report containing the results of the work was published by WRAP in August 2007 [32]. The research evaluated the use of the surface polymerisation technology for the production of recycled rubber particles for use in recycled rubber/virgin rubber blends for retreading truck and earthmover tyres. Specially formulated polymer systems were deposited onto the surface of waste rubber particles using standard rubber industry equipment to create the new compounds. The activated rubber particles were then added to virgin rubber compounds at a high level to produce high-quality vulcanised tread compounds. The truck tyres that were produced from the blends, at both 40 and 50% activated recycled rubber contents, were manufactured by the Tun Abdul Razak Research Centre (TARRC) and tested in their laboratory, and by using road tests. The results obtained were compared to commercially available tyres manufactured using virgin compounds. The earthmover tyres (Figure 6.11) were retreaded by OTR Tyres Ltd of Alfreton and also contained 40 and 50% activated recycled rubber. These tyres were fitted to dump trucks for wear performance trials on working construction sites and quarries. The results that were obtained are summarised below.

Figure 6.11: Surface polymerisation to enable high inclusion rates of rubber crumb into retread compounds for truck and large earthmover tyres. Reproduced with permission from the Waste and Resources Action Programme, Banbury, UK, p.7. ©WRAP (see note with Figure 6.7).

TARRC truck tyres:
- The tests showed impressive results for tyres with 40% recycled content on low- and medium-severity applications (e.g., use of the third axle of a tipper wagon) with a reduction in rolling resistance of almost 20% and an increase in service life of up to 20%. Increasing the recycled content to 50%, however, was found to reduce service life by 34%.
- The results from the high-severity tests (i.e., use on the fourth axle of a tipper wagon) indicated that the 40% tyre, whilst achieving lower rolling resistance, did have a reduction in service life of 25% compared to the controls. This increased to a 50% reduction with the 50% recycled content tyre.

OTR Tyres Ltd earthmover tyres:
- The trial showed that both 40 and 50% recycled content tyres had competitively low wear rates, with the 50% tyre providing a better overall performance than the 40% tyre and achieving a performance equal to the OTR's premium retread tyre, and outperforming a new tyre.
- Overall, the study demonstrated that high percentages of crumb rubber can be treated using the surface polymerisation method and incorporated back into virgin compound with a resulting specification that competes with the market standard products.

Russian workers [33] have used an electromagnetic field to activate the surface of EPDM crumb obtained from waste EPDM products. Having activated the EPDM crumb they then compounded it into EPDM automotive weather strip compounds based on Dutral TER 4038 rubber. Their results showed that the activated powder could be incorporated at levels of up to 20%, in the presence of 1% VTs- 20KP process aid, without causing a significant increase in viscosity, or a reduction in profile stability or tensile strength.

Glogauer [34] has reported on how an innovative atmospheric- pressure plasma technique allows in-line rubber and plastic surface treatment processes to become fully automated with total process control. The new process is claimed to achieve microfine cleaning, high surface activation and nanocoating in a more cost-effective and environmentally friendly way than traditional plasma activation techniques.

Workers at the Chinese university of Yangzhou [35] have used microwaves to modify the surface of waste rubber crumb by devulcanising it and then blending it with NR in various proportions. These mixtures were then vulcanised and the mechanical properties, compression set, swelling behaviour and crosslink density investigated. The results obtained were compared with those of blends that had been prepared using crumb that had not been treated with microwaves.

References

1. M. Massarotto, J.S. Crespo, A.J. Zattera and M. Zeni, *Materials Research*, 2008, **11**, 1, 81.
2. T. Weber, A. Zanchet, R.N. Brandalise, J.S. Crespo and R.C.R. Nunes, *Journal of Elastomers and Plastics*, 2008, **40**, 2, 147.
3. A.N. Gisbert, J.E.C. Amoros, J.L. Martinez and A.M. Garcia, *Polymer Plastics Technology and Engineering*, 2008, **47**, 1–3, 36.
4. *Fine Rubber Powders from Rubber Crumb and Buffings*, Project TYR0009-26, WRAP, Banbury, UK, 2006.
5. M. Orsen, *Rubber Technology International*, 1996, 253.
6. J.A. Shell, *Rubber World*, 2009, **239**, 4, 32.
7. D. Shaw, *European Rubber Journal*, 2010, **192**, 3, 18.
8. D. Shaw, *European Rubber Journal*, 2010, **192**, 6, 9.
9. G. Kostyal in *Proceedings of the Tire Technology Expo 2014*, 11–13th February, Cologne, Germany, 2014.
10. *CWC Best Practices in Scrap* Tires & Rubber Recycling: Ambient versus Cryogenic Grinding, CWC, Seattle, USA,1998http://www.asphaltrubber.org/ari/General_Information/Ambient_vs_Cryogenic_Grinding.pdf
11. R. Kohler and C.G. Astafan in *Proceedings of the ACS Rubber Division Meeting – Fall 1995*, Cleveland, USA, American Chemical Society, Washington, DC, USA, 1995, Paper No.35.
12. R.E. Winters in *Proceedings of the ACS Rubber Division Meeting – Fall 1986*, Atlanta, USA, American Chemical Society, Washington, DC, USA, 1986, Paper No.62.
13. T. Chaubey and H. Arastoopour, *Journal of Applied Polymer Science*, 2011, **119**, 2, 1075.
14. *Scrap Tire News*, 2008, **22**, 3, 6.
15. S.J. Asadauskas, E. Kaminskas, T. Langaitis and A. Jukna in *Proceedings of the Tire Technology Expo 2014*, 11–13th February, Cologne, Germany, 2014.
16. W.J. MacKnight and A.V. Tobolsky in *Polymeric Sulfur and Related Polymers*, Interscience Publishers, New York, NY, USA, 1965.
17. K. Murakami and K. Ono in *Chemorheology of Polymers*, Elsevier, New York, NY, USA, 1979.
18. A. Acetta and J.M. Vergnaud, *Rubber Chemistry and Technology*, 1981, **54**, 302.
19. A. Accetta and J.M. Vergnaud in *Proceedings of the ACS Rubber Division Meeting – Fall 1981*, Cleveland, USA, American Chemical Society, Washington, DC, USA, Paper No.70.
20. A.R. Tripathy, J.E. Morin, D.E. Williams and R.J. Farris in *Proceedings of the ACS Rubber Division Meeting – Fall 2001*, Cleveland, OH, USA, American Chemical Society, Washington, DC, USA, 2001, Paper No.10.
21. J.E. Morin, D.E. Williams and R.J. Farris, *Rubber Chemistry and Technology*, 2002, **75**, 955.
22. D.E. Williams, *Rubber World*, 2002, June, 40.
23. J.E. Morin and R.J. Farris, *Polymeric Materials: Science and Engineering*, 2001, **84**, 1063.
24. J.K. Kim and S.H. Lee, *Journal of Applied Polymer Science*, 2000, **78**, 1573.
25. E. Bilgili, A. Dybek, H. Arastoopour and B. Bernstein in *Proceedings of the ACS Rubber Division Meeting – Fall 2000*, Cincinnati, OH, USA, American Chemical Society, Washington, DC, USA, 2000, Paper No.87.
26. E. Bilgili, A. Dybek, H. Arastoopour and B. Bernstein, *Journal of Elastomers and Plastics*, 2003, **35**, 235.
27. O. Kuznetsova, D. Solomatin, E. Prut and N.N. Semenov in *Proceedings of Advances in Plastics Technology* APT '09, Katowice, Poland, Institute for Engineering of Polymer Materials and Dyes, Gliwice, Poland 2009, p.10.
28. E. Kowalska and S. Pasynkiewicz, *Rubber Fibres Plastics International*, 2010, **5**, 1, 22.

29. R. Burlet and W. Dierkes in *Proceedings of the International Rubber Conference*, Institute of Materials, Manchester, UK, 1996, Paper No.53.
30. W. Dierkes, *European Rubber Journal*, 1996, **178**, 10, 36.
31. F.J. Stark and D. Wagner in *Proceedings of the Rubbercon '95 Conference*, Nordic Council of Rubber Technology, 1995, Paper H1.
32. *Using Waste Tyre Rubber in Retreads at High Inclusion Rates*, Project TYR3-008, WRAP, Banbury, UK, 2007.
33. M.V. Adov, P.L. Krasnov, S.Y. Pichkhidze and L.G. Panova, *International Polymer Science and Technology*, 2010, **37**, 6, T/47–T/49.
34. S. Glogauer, *Rubber and Plastics News*, 2009, **38**, 23, 16.
35. M. Caiyun, L. Ping, M. Limin, H. Hui, Y. Changhao and Z. Ming, *China Synthetic Rubber Industry*, 2010, **33**, 5, 366.

7 Products made from rubber crumb and waste rubber

7.1 Introduction

The option of re-using waste rubber by devulcanising it and then using the devulcanised material to manufacture new rubber products has been covered in Chapters 4 and 5. These two chapters covered waste rubber that had been devulcanised using the wide range of techniques and processes that have been developed deliberately to achieve that goal and also reprocessing techniques that, by involving high levels of shearing (e.g., milling-based techniques), involve at least some devulcanisation by that mechanism.

This section covers the re-use of waste rubber that has been size-reduced by one or more of the processes described in Chapter 6, Section 6.2. For a number of the end-use applications in this section that involve blending or mixing with a new material, better results are usually obtained, irrespective of the matrix type, if the surface of the rubber is activated prior to incorporation by using one of the techniques described in Chapter 6, Section 6.6. As mentioned in Chapter 6, Section 6.6, this improvement in properties occurs due to enhancements in the level of interfacial chemical interactions (e.g., van der Waals forces of attraction) that take place as a result of these treatments.

Once it has been commutated into powder, crumb or shred, there are a number of ways in which waste rubber can be re-used to manufacture new products. For example:

- Blending with thermoplastics to modify properties or produce thermoplastic rubbers.
- Blending with rubber compounds as modifiers or replacement for virgin rubber.
- In construction products, e.g., concrete products, road and footpath surfaces, roofing slates, wood products, insulation materials, civil engineering applications and so on.
- In sports and recreational surfaces.
- Miscellaneous end-uses, e.g., blends with thermosetting resins, as horticultural mulch, in carpet underlay and so on.

The principal ways in which waste rubber can be re-used by incorporating it into this wide range of matrices are covered in Sections 7.2–7.6. The main approach used in these sections is to disseminate the information and describe the work that has been carried out in academia, industry and research organisations to provide recent examples from the published literature. Readers who are interested in a particular area can then obtain an ‘entry point’ into the field by accessing these original papers and research reports.

https://doi.org/10.1515/9783110644142-007

Further information in particular areas can be obtained by using recent reviews. For example, readers who would like to obtain more information on the use of waste rubber in blends with thermoplastics, thermosets and virgin rubber compounds can obtain it in an extensive review that has been produced recently by Karger-Kocsis, Meszaros and Barany [1]. This review also surveyed the methods available to reclaim waste rubber, the surface treatment of rubber particles to improve interfacial adhesion in blends, and the principals underlying the compatibilisation of waste rubber within the host matrix.

Examples of more specific and targeted reviews are provided within the various sections, for example, in Section 7.4.2, where there is a review of the worldwide use of rubber crumb over the past 25 years to modify asphalt [2].

7.2 In blends with thermoplastics

A considerable amount of work has been carried out to establish if it is possible to produce good-quality rubber/thermoplastic blends by incorporating waste rubber crumb, whether surface-activated or not, into thermoplastics matrices. This work has often been carried out using commodity semi-crystalline thermoplastics, such as low-density polyethylene (LDPE), high-density polyethylene (HDPE) and polypropylene (PP), with the aim to produce final products that are either modified (e.g., improvements in impact strength) or have properties that are usually associated with thermoplastic rubbers.

Blends of rubber crumb and thermoplastics can be achieved using either batch processes involving internal mixers, or in continuous processes in which extruders, usually twin-screw compounding extruders, are employed. When extruders are used, there is greater potential to use side feeders to add into the products – before or after the principal blending regions – additives such as compatibilisers, crosslinking agents, antigradants and oils or other additives such as fillers and pigments.

The properties of the resulting blends are dependent on a number of factors:

- The type of ground rubber (e.g., whole tyre, tyre tread or other type, e.g., ethylene-propylene-diene monomer, EPDM).
- The method of size reduction of the crumb (e.g., cryogenic, ambient or water jet).
- The particle size of the ground rubber.
- The type of thermoplastic polymer.
- The proportion of ground rubber used in the blend.
- The compatibilisation technique used to promote interfacial interaction and the resulting degree of compatibilisation.

Murphy and co-workers [3] have looked into the effect on physical properties of blending ultrafine rubber particles into thermoplastics. A large range of blends were produced using a number of recycled thermoplastics and recycled rubbers in a batch process. The influence that the particle size of the rubber, the total amount of recycled rubber, and the degree of compatibilisation had on the physical properties was determined and reported.

It has already been emphasised that achieving a high degree of interfacial interaction is important in optimising the physical, morphological and rheological properties of waste rubber/thermoplastic blends. Zhu and Tzoganakis of Waterloo University [4] investigated the influence that three interfacial agents have on phase compatibility of blends of PP and tyre crumb. The three agents that they used were degraded PP, hydrosilylated PP, and a co-polymer of PP grafted onto styrene-butadiene rubber (SBR). When the resulting compounds were investigated, the interfacial agents were found to be mainly located on the surface of the rubber particles and the degree of improvement noted in the properties was dependent on the functionality of the agents.

The possibility of blending waste injection-moulded PP with waste tyre crumb has been explored by a team drawn from both Loughborough and Moratuwa universities [5]. The two starting materials were blended in a range of different proportions, and samples of each were fully characterised using a range of physical tests and analytical methods. The team was particularly interested in how the blend ratio influenced the crystallinity and phase morphology of the material and how this affected its processibility and mechanical properties.

Work carried out in China and Korea [6] has looked into the role that bitumen and compatibilisers can play in waste tyre rubber/waste PP blends. In addition to assessing the effect of the level of bitumen, the study also investigated the effect that various compatibilisers had on the properties of the final products. The team carried out this assessment using a universal testing machine, a scanning electron microscope (SEM), thermogravimetric analysis (TGA) and a capillary rheometer. The results obtained indicated that the properties of the blends were dependent on the content of the bitumen and the type of compatibiliser.

The possibility of preparing thermoplastic rubbers by blending PP with dynamically vulcanised waste butyl rubber has been reported by a group of Chinese workers [7]. They carried out a series of tests on the blends using a range of techniques to characterise their mechanical, processing and solvent resistance properties.

PP has also been blended with various proportions of recycled nitrile rubber (rNBR) crumb in the presence of a co-agent (trimethylolpropane triacrylate, TMPTA) [8]. The level of TMPTA used was 3 phr based on the amount of nitrile rubber (NBR) and crosslinking of the mixtures achieved using electron-beam irradiation at a level of 40 kGy. The irradiation process was found to improve both the interfacial adhesion between the phases in the blend and their physical properties when these were compared with non-irradiated samples.

A group of workers in Malaysia [9] have compared the properties of PP blended with waste NBR and PP blended with virgin NBR. All of the blends were prepared in a laboratory mixer. Work was done on these blends to determine a wide range of properties and the results obtained showed that the blends prepared with virgin NBR exhibited higher tensile properties and swelling in solvents, but lower thermal stability than blends made with rNBR. It was also found that the morphology of the blends prepared with the rNBR were not as fine as for those from the virgin rubber.

Another group in Malaysia [10] have looked into the viability of using reclaimed rubber from rejected NBR gloves in blends of PP for the production of thermoplastic vulcanisate(s) (TPV). The reclaimed NBR was evaluated as a partial replacement for virgin NBR in the blends in the range 40–70 phr. The TPV were prepared from the starting materials using a Banbury internal mixer, with the vulcanisation achieved using either sulfur or peroxide systems. The TPV compounds had good flow properties and were capable of being injection-moulded. The results obtained on the moulded products using a wide range of tests indicated that the reclaimed NBR could be used to replace up to 70 phr of the virgin NBR without significantly diminishing the final physical properties of the TPV, or its oil resistance.

Ismail and co-workers [11] have reported on the compatibilising effect of epoxy resin on blends of PP and rNBR for the manufacture of thermoplastic rubbers, which exhibit improved mechanical properties. In addition to the effect that the epoxy resin had on the mechanical properties, they also investigated its effect on swelling and morphological properties. The results revealed that the epoxy resin improved the tensile strength, Young's modulus and elongation at break, and reduced swelling behaviour. These improvements were shown by SEM images to be due to an improvement in the adhesion between the rubber and thermoplastic phases in the matrix.

Jyco Sealing Technologies [12] reported on how it can use re-ground waste EPDM in extruded TPV for high-performance automotive weather strip applications. The EPDM material is called JyGreen, requires no re-compounding, and scrap created during its processing into TPV can be recycled. The manufacturing process using the JyGreen material offers environmental benefits in that it is free of volatile organic compound(s) (VOC) and results in less particulate emissions than normally associated with EPDM processing.

A multi-national group of workers in the Far East [13] has contributed to a programme of work that has investigated the effects of virgin EPDM and PP-*g*-maleic anhydride compatibilisers, of various concentrations, on the mechanical properties of thermoplastic blends of PP with waste tyre powder. The workers utilised a design of experiment methodology to optimise the experimental approach in conjunction with a hybrid artificial neural network – a genetic algorithm technique for predicting blend properties. This resulted in an optimised recipe with respect to tensile strength, modulus at 100% elongation, elongation at break, and Shore hardness. The accuracy of this recipe was confirmed by experimental work and led to it being used to manufacture a pad for an automotive brake pedal.

A team of workers from two Brazilian universities [14] have used the collection of statistical and mathematical techniques referred to as response surface methodology (RSM) to investigate mixtures of PP, EPDM and waste tyre rubber. The mixtures were processed using a co-rotating twin extruder, and samples were produced by injection moulding. The physical properties of these samples were determined and their morphology was examined using SEM. Applying RSM enabled the minimum number of experiments to be ascertained in order to optimise the physical properties of a blend produced from the three starting materials.

Montagna and Santana [15] prepared blends of waste tyre crumb and post-consumer PP and high-impact polystyrene and, after preparing samples by injection moulding, studied the effects of rubber concentration and particle size on rheological, physical and mechanical properties. They also investigated the fracture surfaces of the blends after tensile testing using an SEM to characterise the interactions that were occurring between the various phases.

A Malaysian group [16] has prepared products made of latex-modified waste tyre dust blended into PP. The results showed that the tensile properties and swelling resistance of PP-latex-modified waste tyre dust were better than those of PP-unmodified waste tyre dust. The use of the modification was shown to improve the interfacial adhesion between the PP and the dust, improve the dispersion of the dust in the PP matrix, and it did not show any adverse effect on the thermal properties of the product.

Work to investigate the advantages of using ultrafine particles of scrap tyre rubber in blends with plastics and rubbers has also been carried out by McAfee and Grub [17]. These two workers incorporated various amounts of this material into PP, SBR and polychloroprene matrices and investigated their rheological properties using viscosity, torque and tan δ measurements. In this way they obtained data on how the particle size of the rubber, the compatibiliser levels and total rubber content influenced the rheological properties of the compounds.

Workers in India [18] have also blended both waste plastic and waste rubber. In this instance they took scrap computer plastics (e.g., acrylonitrile-butadiene-styrene and styrene-acrylonitrile) and blended them with waste NBR powder. The blends were analysed by TGA in a nitrogen atmosphere and their activation energies of decomposition determined using Friedman and Flynn–Wall–Ozawa methods.

Blends have been prepared from ground tyre rubber and linear low-density polyethylene (LLDPE) at Sichuan University in China [19]. The effect of kneading time and various functional polymer additives on the mechanical properties was documented. These results revealed that the blend that contained a styrene-butadiene-styrene (SBS) functional polymer, and had been prepared using a kneading time of 20 min, gave the best mechanical properties. Analysis of this blend by atomic force microscopy and differential scanning calorimetry (DSC) showed that this was due to a good level of interfacial roughening and interfacial adhesion between the particles and the surrounding LLDPE matrix. The results also showed that it was possible to incorporate as much as 61% of the ground rubber in this blend without compromising its properties too much. Some of the blends displayed only a moderate level of tear strength and elongation at break.

Da Costa and Ramos [20] have prepared LLDPE and EPDM blends, with and without waste tyre rubber, in a single-screw extruder operating with a temperature profile from 180 to 210 °C. They observed that when the EPDM in the blend is partially replaced by the waste tyre rubber, differences in the thermal behaviour of the material and its rheological properties were detected. It was postulated that the reason for this was the possible formation of composite waste rubber/EPDM particles within the rubbery phase of the material.

Hrdlicka and co-workers at the Prague Institute of Chemical Technology [21] prepared blends by compounding LDPE with ground tyre rubber in an internal mixer. They determined the tensile properties, resilience and hardness, and found that they behaved like thermoplastic rubbers. Their work then focused on trying to improve the properties in a variety of ways, including masticating the rubber before blending, replacing some of the waste rubber with virgin EPDM, use of compatibilisers, and carrying out partial crosslinking using peroxides or sulfur.

The properties of blends of ground tyre waste and thermoplastics can be altered by irradiating them with energy sources, such as gamma rays and ultraviolet (UV) radiation, to form crosslinks within the matrix. Treating blends of thermoplastic and rubber in this way to generate crosslinks will alter a wide range of physical properties. A group in Egypt [22] have carried out a study on how these two crosslinking methods change a number of the mechanical, physical, morphological and thermal properties of blends of EPDM, ground tyre rubber and HDPE. They also looked into the influence that the ratio of these three components had on these properties.

A solid-state mechanochemical milling process has been developed that combines both the devulcanisation of tyre waste and its mixing into an HDPE matrix [23]. The resulting product was also dynamically vulcanised within the mill, resulting in a thermoplastic elastomer (TPE) that SEM analysis showed to have good interfacial compatibility between the rubber and thermoplastic phases.

A group of workers in the Ukraine [24] has studied the effects of multiple processing on the properties of thermoplastic rubbers produced using waste-type rubber, recycled HDPE and EPDM rubber. Results obtained by DSC and dynamic mechanical analysis (DMA) showed that, as the number of processing cycles increased, the phase separation between the amorphous and crystalline phases in the TPE decreased. Increases in the crosslink density that resulted from the multiple processing did not appear to have a significant effect on the physical properties of these TPE.

Yagneswaran and co-workers [25] incorporated mechanically devulcanised waste tyre powder into a vinyl ester resin matrix at various levels of addition and characterised the resulting materials using physical tests, DSC and TGA. As the level of rubber powder increased the composites were found to have reduced heat stability and flexural strength, but increased flexural modulus.

7.3 In blends with rubber compounds

At the start of this section it is important to reiterate the difference between this section and Chapter 5. Chapter 5 described the work carried out to produce new compounds and products by blending devulcanised rubber with virgin rubber compounds. That chapter and this section have, as a common link, the fact that the starting point for devulcanisation processes is very often rubber crumb. In the majority of devulcanisation processes, however, the crumb will lose its original morphological structure during the process and a new material is produced that, ideally, closely resembles the original virgin rubber.

The investigations and studies presented in this section are concerned with the blending of waste rubber powder or crumb, which has not been through any devulcanisation processes of the type described in Chapter 4, into rubber compounds to produce new products. The rubber crumb in question may or may not have been surface-activated by one of the processes described in Chapter 6, Section 6.6. The properties of the resulting blends will depend upon whether this activation has taken place or not, as well as upon some of the variables already listed at the start of Section 7.2, namely the origin and type of rubber crumb, the proportion of rubber crumb in the blend, and the type of rubber into which it is being blended. As with all rubber compounds, the final properties will also depend to a large degree on the type of cure system, the crosslink density of the rubber, and the presence and level of key ingredients such as fillers and plasticisers.

One of the goals of the tyre industry for many years has been to try and re-use as much waste tyre rubber as possible, without compromising safety and performance, in the manufacture of new tyres. Progress has been made and this has been demonstrated by a number of academic and industrial workers. For example, Li and Zhang [26] of the Qingdao Nexen Tire Company in China have shown that it is possible to incorporate 5 phr of low-odour, fine tyre tread crumb into a new tyre tread and, by increasing the level of curing agents in the compound, the physical properties were found to meet the requirements of tread compounds. In particular, the endurance and high-speed performance of the tread were found to be good. Another benefit was that the cost of production was reduced by the use of the waste rubber.

Guzman and co-workers [27] investigated whether it is possible to use waste tyre crumb as a replacement for zinc oxide as an activator in the sulfur vulcanisation of natural rubber (NR). They used the unsaturated organic compound squalene as a model compound for NR in their work, and followed the course of the vulcanisation reaction using the analytical technique high-performance liquid chromatography. The results confirmed that waste rubber crumb was an alternative to zinc oxide as an activator in the curing of NR compounds by sulfur-based cure systems.

Chandra and Pillai [28] used a low-molecular-weight recycled rubber as a processing aid in a number of tyre tread formulations and compared its performance with other commercial process aids. This process aid (Ecoreen T) is shown to function as well as or even better than depolymerised NR or similar process aids. The same workers also reported on the use of new products with low viscosity that are produced by blending low-molecular-weight virgin polymers with recycled butyl rubber [29]. When these products are added in small doses of between 5 and 20% as a replacement for virgin rubber in formulations, it helps to reduce the shrinkage of calendered rubber sheets, reduces mould shrinkage, and enhances some physical properties, in particular resistance to crack growth on flexing. The products were found to be useful in products that require high resistance to flex cracking such as curing bladders and liners and they take part in the vulcanisation chemistry to cure into the host matrix. Other useful properties that are claimed for the products are that they are useful as an impact modifier in polyamides and can be used in the preparation of TPE.

As has been shown in Chapter 6, in addition to crumb from waste tyres, it is possible to commercially source crumb from a number of other rubber types, e.g., fluorocarbon rubber. Shell [30] has carried out a study whereby micropowder from vulcanised fluorocarbon, silicone and nitrile wire insulation rubber has been put back into wire compounds using an open mill. The loadings chosen were those that maximised the performance and economic efficiency of the formulations.

Workers in Brazil [31] have investigated the properties of revulcanised SBR that has been recovered from scrap rubber. Their work involved compounding the recovered SBR with an aromatic oil and different conventional rubber cure systems. The curing behaviour of these compounds was then evaluated using a rubber process analyser.

Waste rubber crumb from a number of diene rubbers such as SBR, butadiene rubber (BR) and polyisoprene has been compounded with cheap mineral fillers and additives, such as monoethanolamine, by Ushmarin and co-workers [32] at Chuvash State University and used to manufacture general rubber goods such as car mats and 'sleeping policemen' for railway lines. They found that the use of a two stage mixing process in an internal mixer produced materials with the best physical and mechanical properties.

The use of up to 15 phr of recycled rubber powder and reclaimed rubber to replace some of the BR and NR in rubber compounds has shown that it increased the rate of cure of the compounds [33]. This enabled the quantity of cure ingredients in the compounds (e.g., accelerators or zinc oxide) to be significantly reduced, resulting in an additional cost saving. In addition to monitoring the influence of the waste rubber on curing characteristics, its effect on strength properties, resilience, tear strength and crack growth were also assessed.

Baeta and co-workers [34] have investigated the use of waste SBR crumb as a filler for NBR compounds. Their paper describes the effect of the particle size of the crumb and the loading on a range of properties such as cure behaviour, mechanical strength, swelling in solvents and crosslink density.

Three Malaysian workers [35] have carried out an investigation into the effects that different particle sizes of rNBR have on the curing properties, and the mechanical and morphological properties when it is blended with SBR at different blend ratios. Three different particle sizes were used: 117–334 µm, 0.85–15.0 mm and 10–19 cm. The SBR/rNBR blend ratios, prepared using a two-roll mill at room temperature, were 95:5, 85:15, 75:25, 65:35 and 50:50. Testing of the blends showed that the scorch time and cure time decreased with increasing rNBR content as well as decreasing particle size of the rNBR. This was thought to be due to the presence of substances such as unreacted curatives in the rNBR. The blends with the smallest particle-sized rNBR showed the minimum torque values and gave the best processing characteristics, and also displayed the best physical and mechanical properties, but had inferior resilience. An analysis of these blends with an SEM showed that the improvement in properties was due to a better interaction between the rNBR and SBR fractions within the rubber matrix.

Noriman and co-workers [36] prepared blends of SBR and rNBR, some of which contained the polyfunctional monomer TMPTA, and then irradiated them with electron-beam irradiation. The content of the rNBR in the blends varied in content up to at least 50 phr and, above 15 phr, irradiation was found to increase a number of properties including crosslink density, stress at 100% elongation, hardness and tensile strength. The elongation at break and resilience were found to decrease, however. The thermal stability of the blends was also found to be reduced by irradiation. The same group of workers also studied the effects of epoxidised natural rubber (ENR-50) as a compatibiliser on the properties of blends of SBR and rNBR prepared using a two-roll mill [37]. Five different blend ratios were used, with the ENR-50 content fixed at 10 phr. A wide range of tests were performed on the blends, with the results showing that the ENR-50 did improve the compatibility of the two rubbers. It also resulted in improvements in the tensile strength, but a reduction in the elongation at break and the rebound resilience. They also compared the curing properties, mechanical properties, crosslink density and resilience of virgin SBR/virgin NBR blends with virgin SBR/rNBR blends [38]. The programme of work included examining the fatigue fracture surfaces using an SEM to determine the quality of the interfacial blend interactions and look for any microdefects in the matrix.

Noriman and co-workers [39] investigated the influence that rNBR has on the weathering performance of blends of rNBR and SBR. The workers prepared the blends using rNBR in three different forms, designated S1 (117–334 µm particle size), S2 (0.85–15 mm particle size) and S3 (sheet form). They vulcanised the blended compounds and exposed test pieces to natural weathering conditions for either three or six months. The results showed that the best retention of physical properties, such as tensile strength, and general resistance to weathering, measured by the increase in concentration of UV-induced oxygenated structures, was exhibited by the blends that contained the rNBR in its finest particle-sized form.

The potential for using reclaimed NBR gloves as a partial replacement for virgin NBR (at levels of up 70 phr) in NBR compounds has been discussed by Hassim and Ahmad [40]. The results indicated that the recycled gloves could be used as a replacement for virgin NBR at levels of up to 70 phr without having an adverse effect on the physical properties of the vulcanisates or their oil resistance. Use of the recycled gloves also offered the possibility of a significant reduction in manufacturing costs.

A Russian project [41] has investigated the possibility of using activated waste EPDM powder in virgin Dutral TER 4038-based EPDM rubber compounds for use as weather strip in automotive vehicles. The rubber powder was activated in an electromagnetic field and the data obtained on the prepared samples showed that it could be added at a level of up to 20% without any significant increase in the viscosity of the EPDM mixes. The small increase that did result could be prevented by the use of a VTs-20KP processing aid at a level of 1%. The test data obtained on samples prepared at the 20% addition level showed that the deformation and strength properties were the same as Dutral TER 4038 compound that did not contain any waste EPDM.

Adov and co-workers [42] have shown that activated fine waste EPDM rubber powder can be incorporated into EPDM compounds at levels of up to 20% without altering the physical properties. A slight increase in Mooney viscosity was found, but this was compensated for by the use of a process aid.

Another study that concerned the use of an activated rubber crumb was also performed by Adov and co-workers [43]. They introduced a finely ground activated polychloroprene powder into a virgin chloroprene compound and made a quantitative assessment of the Payne effect in the vulcanisate. They produced data that showed the dependence of the real and imaginary parts of the dynamic shear modulus, and the dependence of the mechanical loss tangent on the logarithm of the strain amplitude for vulcanisates containing different contents of activated powder. Their results indicated that the introduction of the activated chloroprene powder into the vulcanisate promoted an improvement in the interaction of the rubber with the filler present in the compound, which should lead to an increase in the level of service properties for the products and open up a route for the effective re-use of the rubber waste.

Blending waste ground rubber with elastomeric materials such as ethylene-vinyl acetate (EVA) copolymer can improve the damping properties of the resulting products. Such an investigation was carried out by a team of French workers [44]. To evaluate the influence of both the ground rubber and the porosity in the ground rubber/EVA composite, three samples were obtained by injection moulding: the EVA alone, the ground rubber/EVA composite and a porous ground rubber/EVA composite. Following analysis by DMA, the Young's modulus and loss factor of these three materials were estimated using the frequency–temperature equivalence introduced by Williams–Landel–Ferry. This method showed that, in a low-frequency bandwidth, the loss factor was improved by the incorporation of the ground rubber into the EVA matrix. The alpha-relaxation work showed a lower activation energy for both the ground rubber filled composites, leading to the conclusion that the mobility of the polymer chains had improved. Also, impact testing carried out using a drop weight test showed better impact energy absorption for the ground rubber filled composites.

7.4 Use in construction products

Because of the possibilities for utilising large amounts of waste rubber and the potential environmental, economic and technical advantages that using rubber in products such as concrete could yield, construction products is an area that has also seen a significant amount of research. Waste rubber is also a promising material for use in the construction industry due to its low density, elasticity, energy absorption and sound- and heat-insulating properties.

7.4.1 Concrete products

The use of waste rubber and other polymers such as polyethylene terephthalate (PET, from bottles) to modify concrete is an active area of research. One of the attractions of this recycling is the potential to use up large amounts of waste material. A recent paper [45] by Pacheco-Torgal and co-workers reviewed the research carried out in this area, and discussed the influence that the size of the rubber particles and the total amount of waste rubber or PET have on the properties of both wet and hardened concrete.

When it is used in concrete, rubber crumb is either a partial or total replacement for conventional aggregates. A complete replacement of the aggregate by crumb gives a product that is regarded as plain rubberised concrete and this can be typically used for non-structural applications, where its relative weakness does not present any problems. An alternative product, with partial replacement, is referred to as self-compacting rubberised concrete (SCRC) and a paper was published in 2010 [46] that summarised and compared the results obtained in over 70 independent studies on this material. These studies focused on the properties of both fresh and hardened SCRC products and described the key characteristics of the two forms.

In 2005, WRAP funded a project [47] that conducted trials to explore the possibilities of introducing waste tyre rubber chips, granulate or powder into a range of concrete blocks as a replacement for primary aggregates (Figure 7.1). The project was carried out for WRAP by Brand & Rae Limited and was conducted along similar lines to a project worked on by WRAP and Brand & Rae to investigate the use of crushed glass as an aggregate replacement in concrete blocks [48]. The work consisted of three phases:

- *Phase I*: Production of a report summarising existing research and material available.

Figure 7.1: Lightweight construction blocks for the construction industry manufactured using waste tyre rubber as a partial replacement for aggregate. Reproduced with permission from the WRAP Quality Protocol, Waste and Resources Action Programme (WRAP), Banbury, UK. ©WRAP.
Note: WRAP does not warrant that the functions contained in the material contained in this Site will be uninterrupted or error free, that defects will be corrected, or that this Site or the server that makes it available are free of viruses or bugs or represents the full functionality, accuracy, and reliability of the materials.

- *Phase II*: A range of concrete mixes were produced in the laboratory and the test samples used to assess the effect that the waste rubber had on the concrete and how this related to the strength of the overall product. The results obtained were used to make recommendations for Phase III of the project, and they showed that once more than 10% of the aggregate was replaced with rubber in the concrete, its strength fell away rapidly. The addition of more cement to offset this reduction did not give a sufficient improvement in strength to justify the cost. It was therefore concluded by the workers that the optimum level of waste rubber in the concrete blocks was between 5 and 10%.
- *Phase III*: Two types of concrete block were produced with rubber replacing aggregate in the range 5–10% to find the optimum amount. The optimum amount was found to be 4%. A full-scale production trial was conducted using this level, with the manufacture of approximately 6,000 blocks containing 30 tonnes of tyre rubber.

The products produced in Phase III were subjected to a full range of tests carried out at Brand & Rae and at an approved laboratory. The results showed that there was some reduction in the strength of the blocks (1.25 N/mm^2) compared to the control product that did not contain any rubber, but that the presence of the rubber did improve the thermal conductivity, reduced the level of water absorption, and increased the density by 105 kg/m^3, yielding an improvement in noise attenuation.

El-Gammal and co-workers [49] investigated the density and compressive strength of concrete that contained waste tyre rubber at various levels. The waste rubber had been used to replace equivalent amounts of fine and coarse aggregate in the test mixes. The results showed that, although there was a significant reduction in the compressive strength of the rubber modified concrete, the products demonstrated a ductile, plastic mode of failure as opposed to the brittle failure common with standard concrete.

A team from the Universite de Picardie Jules Verne [50] investigated the thermal conductivity of a lightweight construction material containing rubber waste particles. Measurements were carried out in a dry state using a transient plane source technique. To determine the effect that varying the quantity of rubber particles had on the thermal conductivity of the cement, recipes containing 10, 20, 30, 40 and 50% of rubber, as cement replacement, were used. The results revealed that the addition of rubber particles reduced the unit weight of the material and improved its thermal conductivity. In addition, based on the self-consistent method and assuming that the material is a tri-composite consisting of air, rubber particles and cement, the thermal conductivity of it has been predicted as a function of the dry unit weight and various formulations using the autocoherent homogenisation model. This model, which showed a good degree of correlation with the experimental data, required the thermal conductivity of the rubber particles themselves, which was experimentally determined using the Horai and Simmons technique.

Jevtic and co-workers [51] looked into the possibility of using waste rubber crumb to replace some of the natural river aggregate that is used in cement mortar. They carried out a range of tests on both the freshly prepared cement mortar and samples in the hardened state. With regard to the work on the fresh mortar they determined density, consistency and volume of entrapped air. In the hardened state they measured density, mechanical properties, modulus of elasticity, adhesion to concrete substrate, water absorption, freeze–thaw resistance and ultrasonic pulse velocity. An examination of the results revealed that the rubber can be successfully used as partial replacement for the aggregate, that the properties of the resulting materials are dependent on the level of replacement, and that the best results were apparent in the freeze–thaw resistance data.

7.4.2 Bitumen and asphalt products

Another area of research involves the use of crumb in bitumen-based products. An Iranian team have looked at this [52] but also investigated the possibility of using the natural bitumen that is found in several regions of Iran, as a modifier for bitumen. The aim of the study was to look into the possibility of obtaining penetration-grade modified bitumen from vacuum bottom (VB) residue by adding various recycled polymers and non-polymeric materials without air blowing the VB residue. Of the waste polymers used in the study, in addition to crumb rubber, recycled waste latex and waste polyethylene (PE) were also included. A large number of different blends were prepared and the workers found that a combination of 10% natural bitumen and 5% recycled latex with VB residue gave the best properties. Problems with a decrease in ductility were encountered with the waste crumb rubber if it was too coarse and poorly dispersed within the bitumen matrix.

The results of another study that looked into the effects of modifying bitumen with a range of recycled polymers have been published by Shakhnazarli and co-workers [53]. The waste rubber that was used was SBR, and the other polymers included polyamide fibres and recycled PE. A range of properties were evaluated, including extensibility, softening point and adhesion.

At the *4th Rubber Modified Asphalt Conference* held in Akron, OH, USA in 2009, a wide-ranging paper was presented by Baumgartner [54] that addressed topics such as modified asphalt formulations, optimisation of the process temperature, and the ground tyre rubber loadings and particle size optimisation. The paper showed that rubber crumb produced from whole tyres contains around 30% reactive material for asphalt modification and that the asphalt source and chemistry directly influence the rubber loading and the final properties of the product. The processing time and temperature are also very important, as is the particle size of the crumb, which affects the efficiency of modification and the long-term performance.

In March 2008 WRAP published the results of a review carried out to investigate the worldwide use of crumb-modified asphalt over the previous 25 years [2]. The work was conducted for WRAP by Scott Wilson and involved a desk study of published information from 1983 to 2008 on the use of rubberised asphalt (RA) worldwide and the results of communications with relevant industrial experts on the practical issues associated with the use of RA in their countries. There are two main processes for using rubber in asphalt (i.e., the dry and wet processes) and this review concentrated on the use of the wet process, as this is the method to produce rubberised bitumen. Rubberised bitumen can be used in four different areas of application:
- Spray seals
- Stress-alleviating membrane
- Stress-alleviating membrane interlayer
- Asphalt mixture

Some of the other salient findings of the review, which tended to focus on the asphalt mixture applications where rubber had been added at levels of at least 15% by weight, are summarised as follows:
- RA has been in use in the USA since at least the early 1960s. It was initially used in Arizona, but has since been used more widely in four states in the USA and elsewhere, including South Africa, Europe and Australasia.

- Three types of RA have been used traditionally: porous friction course, and gap and open graded asphalt concretes.
- Some of the numerous advantages claimed for using RA material are improved durability and resistance to age-hardening; improved resistance to surface-initiated and fatigue/reflection cracking; reduced temperature susceptibility; lower maintenance costs due to improved durability and performance; and better aggregate retention (i.e., resistance to fretting).
- Some countries only specify open and gap graded RA course mixtures with a binder content of 7% or more due to some workability issues reported with the dense graded asphalt mixtures that have lower binder contents of around 4.6%.
- Construction processes may be more challenging. For example, gap and open graded RA has to be compacted at higher temperatures than conventional dense graded asphalt because the rubber stiffens the binders. The RA mixtures can also be difficult to hand work due to this stiffening, and the construction must be completed 48 h after mixing. Also, construction in cold temperatures, in the wet, and over pavements with severe cracks can cause problems.

Overall the review's findings suggested that RA enhances the performance of asphalt pavements and that the target level of performance could be delivered if good-quality production control is applied throughout the process of mixing and constructing the RA layer.

The most beneficial applications were found by the review to be thin asphalt friction course overlays over portland cement concrete or aged asphalt pavements, to provide quieter and safer ride characteristics (e.g., skid resistance) for the travelling public, or in the spray applied membranes used as interlayers or surface treatments. However, to provide the improvement in pavement performance, the asphalt mixtures must be properly selected, designed, produced and constructed. Important criteria also include the structure of the pavement and adequate drainage. This WRAP review report is associated with another WRAP report, also prepared by Scott Walker, which provides more specific information on the properties of RA for use in the UK [55].

An Egyptian group in Alexandria [56] conducted studies into the effect of incorporating different levels of devulcanised crumb rubber into two different types of asphalt. They used the devulcanised rubber at two different levels (designated low and high), and the two types of asphalt were fluxing propane precipitated asphalt and blown asphalt. The physical-chemical properties, such as softening point and viscosity, of the resulting products were investigated, and the results obtained discussed in terms of the suitability of the products for end-use.

A team from Iran [57] looked into re-using waste products from SBR and HDPE manufacturing plants as modifiers for asphalt derived from the vacuum bottom residue in crude oil production. In addition to the waste SBR and PE, they also included *N*-methyl pyrrolidone (as an extender) in the study. Penetration, softening point and Frass tests were conducted on the vacuum bottom residue, and Marshal tests were conducted on the polymer-modified hot mixed asphalt. They found that the optimised polymer-modified hot mixed asphalt exhibited superior strength and lower flow compared to standard 60/70 penetration-grade bitumen.

An example of waste rubber powder being used to produce a new road surface on a public road was reported in 2012 [58]. In that article it was reported that Breedon Aggregates, a UK-based company, had resurfaced a short stretch of a dual carriageway in the UK using asphalt that contained recycled ground rubber. The rubber crumb used had been obtained from a Danish company called Genan, which produces Road+, a product specifically designed for use in the surfacing of roads. Road+ consists of rubber powder and a second component called Vestenamer, which is a semicrystalline polyoctenamer product by Evonik. Laboratory tests and field road studies have shown that a standard bitumen modified with 10% Road+ as a minimum performs at the same level as a similar 4% SBS-modified bitumen. For several of the properties, for example the ability to prevent rutting, the product modified with Road+ performs better.

A case study has been published by WRAP [59] that describes how a cold micro-asphalt resurfacing product (Rejuvoflex, manufactured by JPCS) containing 40–50% recycled materials, including around 5% by weight 1–2 mm waste crumb rubber, was successfully used to resurface two areas of pathways (a total of almost 4,000 m^2) during March 2006 in the Walsall area of the UK (Figure 7.2). In addition to Rejuvoflex, which uses about 1 tonne of tyres per 1,000 m^2 of laid micro-asphalt, JPCS also produces Rejuvomak, a two-layer product that

Figure 7.2: Pathways in Walsall in the UK have been resurfaced using a micro-asphalt product (Rejuvoflex) that contains waste tyre crumb. Reproduced with permission from the WRAP Quality Protocol, Waste and Resources Action Programme, Banbury, UK. ©WRAP (see note with Figure 7.1).

includes a specialised underlayer, and this combination can utilise almost 3 tonnes of tyres per 1,000 m^2 of laid surface. The WRAP case study describes the technical, health and safety, cost and environmental benefits of using products such as Rejuvoflex.

In 2008, crumb from more than 10,500 scrap tyres was used in an asphalt formulation with rock chips to resurface 38 segments of streets in Suisan City, California, USA [60]. The $1.2 million project involved laying the RA and then, two weeks later, applying a slurry seal that cured into a smooth surface. Using the combination of RA and the slurry seal is thought to add 10 years of useful life to the street surface, as it is more durable that traditional asphalt and fills cracks in the existing asphalt, helping to keep water from the road base. The combination also costs less per unit area than standard asphalt and so enables more streets to be resurfaced with the same budget. Similar projects have also been undertaken in Canada. In 2008, 130,000 waste tyres were also used for RA highway construction projects worth more than $11 million in Saskatchewan [61]. It was estimated that the work used around 1,300 tyres per lane per kilometre. The tyres were collected, processed and recycled through a non-profit, non-government programme run by the Saskatchewan Scrap Tire Corporation.

7.4.3 Gypsum and screed products

Gypsum plaster can cause problems by cracking in the first hours after hardening due to a lack of elasticity, and workers in Spain have evaluated the use of waste tyre rubber to reduce the possibility of this occurring by reducing the elasticity modulus of the set plaster [62]. Their study analysed the mechanical behaviour of gypsum plaster that contained rubber particles of three different sizes (0–1, 1–2 and 2–4 mm) at three different volume fractions (1, 3 and 5%). The results were compared with those for gypsum without any rubber modifier.

WRAP [63] has also funded a project that uses rubber from waste tyres to product new flooring screed and plaster products. The products are generically known as Rubbercrete, with the plaster products containing 70% rubber crumb and the flooring screed 30% crumb. Rubbercrete is also available in a rapid set mix design, which has a crumb content of 35% and dries within two to three hours.

7.4.4 Roofing products

Roofing tiles, slates and membranes are an end-use for rubber crumb and recycled rubber that has been explored by a number of companies and organisations [64]. The WRAP organisation in the UK sponsored a project to look into the potential for using waste rubber crumb in replica roofing slates (Figure 7.3). The research project

Figure 7.3: Synthetic roof slate that contains ultra-fine waste car tyre rubber. Reproduced with permission from the WRAP Quality Protocol, Waste and Resources Action Programme, Banbury, UK. ©WRAP (see note with Figure 7.1).

was undertaken by Queen Mary University in London, Airport Business Centre and Crumb Rubber Ltd and evaluated the use of ultra-fine 80 mesh rubber powder from waste car tyres in a recycled PP matrix for the production of replica roofing slates [65]. Initially, a replica slate from an American supplier was obtained and reverse engineered to establish its composition and properties. This slate was found to contain 17–25% of rubber in a PE matrix and had a tensile modulus of around 850 MPa and a tensile strength at break of 14 MPa. Laboratory trials showed that two blends, of PP and 40% and 60% ultrafine rubber powder, provided properties close to the American slate. Further work then looked at a number of variables, such as the amount of rubber (varied from 20 to 60%), the effect of the melt-flow index (MFI) of the recycled PP (resins with values of 1.2, 9 and 30 were used), the use of compatibilisers (e.g., PP-graft-maleic anhydride), the use of wood flour, micaceous china clay and glass fibres as stiffeners, and virgin HDPE as an alternative to PP. The results showed that PP with an MFI of 1.2 gave the best results and that rubber loadings of between 20 and 30% met the benchmark properties that were adopted for the trial. However, a loading of 40% showed adequate performance to warrant further commercial development, and this level and the highest used in this study (i.e., 60%) were chosen for the production moulding trials that followed this laboratory research and development (R&D) project. None of the materials evaluated as stiffening agents was found to work, including the compatibiliser, but it was found that the slates could be produced with a good-quality finish by either injection moulding or compression moulding. Due to its ability to produce slates more efficiently, injection moulding was chosen for the moulding trials, and a 330 tonne moulding machine was found to be successful in the production moulding trial of both the 40 and 60% rubber modified slates of dimensions 500 × 250 × 5 mm. Initial estimates to come out of the project are that, if the product passes all the certifications need for the UK, it could supply 5% of the English slate roof market, equating to 3,500–5,000 tonnes of rubber per year, or around 5,700–6,200 tonnes of used car tyres per year.

A high-tech method of using waste rubber crumb to manufacture rubber/plastic composite water-repellent sheets that have the potential to be used in the building industry has been reported by a group of Israeli and American workers from both industry and academia [66]. The waste rubber is devulcanised and then remoulded between a PE film and a specialised crimped film to form sandwich sheets that have a high degree of hydrophobicity. Contact angle measurements by the researchers showed these sheets to have contact angles as high as 155°.

An environmentally friendly commercial roofing system has been commercially developed by West Development Group [67]. The roof comprises a soy-based spray polyurethane (PU) foam with a sprayable solvent-free silicone top coat containing recycled EPDM granules forming a water-resistant, seamless, self-flashing seal.

7.4.5 Sound and vibration insulation products

An area that has seen a lot of interest is in the use of waste rubber to produce sound insulation materials. Zhao and co-workers [68] have investigated the properties of waste tyre rubber/wood composites for sound insulation applications. They used a four-microphone method to measure and compare the sound transmission losses of three different composite panels: the waste tyre/wood material, a commercial wooden floorboard, and a commercial wood-based particle board. The waste tyre rubber/wood product was manufactured in the laboratory with commercial urea-formaldehyde and a PU adhesive. The results obtained indicated that the sound insulation properties of the waste rubber/wood material were better than the other two products. Also, its insulation properties were significantly affected by the amount of rubber crumb and the amount of PU adhesive used, the performance improving as the quantity of both ingredients was increased. In the case of the PU adhesive, this was due to the formation of more complete rubber–adhesive interfaces within the material.

The use of ground tyre rubber from heavy-duty tyres to produce a sound-absorbing material has been discussed by Maderuelo-Sanz and co-workers [69]. They produced a novel porous material by blending the ground rubber with PU resin, and assessed its acoustic performance according to International Organization for Standardization (ISO) 10,534–2:1998, as the proportion of rubber to resin was varied, and the particle size of the ground rubber was altered. The results showed that a high sound-absorption performance could be achieved even when the material was produced as a relatively thin sheet. The re-use of waste rubber to manufacture sound-proofing materials has also been investigated by Borlea and co-workers in Romania [70]. They manufactured the products by incorporating waste rubber particles, in combination with other materials such as cork and fibres, into a PU binder, and assessed their sound-absorbing properties by measuring their sound absorption coefficient, as the frequency of the sound energy was varied, using impedance tubes.

Another of WRAP's waste tyre projects [71] also looked into the use of rubber crumb and shred obtained from waste truck tyres to produce a range of items that act as barriers to noise, vibration and impact (Figure 7.4). In these instances the rubber products were encapsulated by mixing with hot PU and then moulded by Island Leisure Products (ILP), based in Sittingbourne, Kent. ILP manufactures pre-moulded rubber products from used tyre crumb and, in total, they use 1,500 tonnes per year of shredded and crumbed truck tyres in their range of moulded products. For this WRAP project they used a mix of 1 mm rubber crumb from truck tyre casings and 5 mm shreds from truck tyre buffings. Hot PU was added to these as a binder and the mixture stirred until it was homogeneous. The hardness and physical properties of the final product are affected by factors such as the mixing speed and the proportion of rubber present. For example, if a softer product is required, the amount of rubber is increased. The rubber/PU mixture is poured into a customised mould, which is heated to 90 °C and then compressed and cured for 5–8 min depending on the thickness of the product. To obtain a change in colour, dyes can be added at the mixing stage and, to reinforce the product or change its

Figure 7.4: Moulded barrier products have been manufactured using rubber crumb and shred from used truck tyres. Reproduced with permission from the WRAP Quality Protocol, Waste and Resources Action Programme, Banbury, UK. ©WRAP (see note with Figure 7.1).

appearance, steel or coloured rubber inlays can be added at the moulding stage. A number of products can be successfully manufactured by ILP in this way, including sound and vibration dampeners, rubber mats, rubber transportation wedges, and steel-reinforced rubber traffic posts.

7.4.6 Miscellaneous products for construction

The use of rubber crumb in adhesives is not new, with records of such applications dating back to at least 1927, but it was not until the 1960s that rubber-containing cementitious tile adhesives were developed, with the rubber being used as a replacement for sand.

This provides an advantage in that sand-containing adhesives are rigid and not suitable for certain applications such as fixing rigid floor tiles to flexible wooden floors, but the use of waste tyre crumb in conventional cementitious adhesives can provide the desired combination of elastomeric bond strength and adhesive flexibility. This type of application for waste rubber is attractive as the demand for rubber-containing adhesives is increasing by an estimated 10% per year due to a combination of product substitution and new product and market development. WRAP and Building Adhesives Ltd, the largest manufacturer in the UK of domestic do-it-yourself (DIY) and professional adhesives that contain between 30 and 60% waste tyre rubber, have collaborated in a WRAP Waste Tyres Case Study [72] addressing the technical, cost and environmental benefits of these products (Figure 7.5).

Waste tyre-derived aggregates, in the form of 50–300 mm shreds, are used routinely in major road, highway and bridge construction projects as lightweight fill and 'compressible inclusions' behind bridge abutments to lessen lateral pressures against the abutment. WRAP have looked at the potential for extending the use of this material, and other waste tyre materials, in commercial and residential construction work [73]. This research was carried out from 15th November 2005 to 23rd February 2006, and describes the benefits of waste tyre-derived rubber products, aggregate and sheet in engineered drainage systems, as a compressible layer to absorb ground movements, and as a noise- and vibration-absorbing medium. One of the precedents for this investigation is that, in the USA, tyre-derived aggregate is used in many of the federal States as a substitute for newly quarried crushed rock in septic system drainage designs and as backfill against foundations and basements because of two large benefits: it is free draining, and it acts against a foundation or basement wall just like it does against a bridge abutment, in that it absorbs lateral pressures, thereby reducing the stress on the structure, allowing substantial savings in reinforcing steel and concrete thickness. Another benefit of rubber's compressibility is that it is useful in buffering substances and superstructures against land movement. A final advantage for these sorts of end-

Figure 7.5: Dunlop^{TM} Tile-on-Wood Rapid adhesive is manufactured using waste tyre crumb. Reproduced with permission from the WRAP Quality Protocol, Waste and Resources Action Programme, Banbury, UK. ©WRAP (see note with Figure 7.1).

use is that waste tyre-derived rubber sheets are important elements in modular building system design as they act as acoustic and vibration barriers in modern construction techniques.

Bales of waste tyres have been used for many years in large construction projects, and this method of re-use was the subject of a report published by the UK Department of the Environment, Food and Rural Affairs (DEFRA) in 2001 [74]. This report reviewed the international literature that was available on baled tyres, in particular the long-term behaviour and stability of the tyres and the metal that binds them together, when used in marine environments (e.g., artificial reefs), watery environments (e.g., flood defence or river management) or acid peat bogs, in order to assess their environmental impact. The scope of the review also included examining baled tyres encased in concrete. As dictated by the waste protocols, tyres may be baled together for storage, but they remain designated as waste until used for construction work.

Shredded tyres can be used in the construction of landfill sites as a replacement for other construction materials. The areas that they are used in include lightweight backfill in gas-venting systems, leachate collection systems and operational liners. Shredded waste tyres can also be used to cap, close or daily cover landfill sites [75]. Their use as a backfill material and cover material can be more cost-effective than the use of other fill materials, as they can be shredded on site rather than being transported in for that particular purpose.

7.5 Use in sports and other recreational surfaces

Rubber crumb can be used very successfully in the manufacture of sports surfaces such as running tracks in sports stadiums and artificial turf for a variety of uses. Commercial examples of such products include a track marketed by Conica, a division of BASF, as Conipur SW, which is a sandwich system comprising an elasticated layer of recycled rubber granules and a PU binder that is applied in situ to an asphalt substrate with built-in finishers [76].

An example of an artificial turf product is one marketed by Mondo as Mondoturf. This is made up of layers comprising a sheet of PE that is placed directly on the soil, a prefabricated elastic mat, the turf surface and an Ecofill infill. About 70% of this system is recycled material, with the elastic mat being derived from waste tyres and the plastic membranes from waste bottles [76].

The use of tyre crumb in the manufacture of sports surfaces can raise questions related to health and safety issues. For example, synthetic turf composed of a layer of tyre crumb and ribbons of PE fibres has been installed in many playing fields in the New York City area in the USA [77]. Concerns were raised that the VOC present in the rubber could cause problems when released into the breathing zone of the players, particularly on hot days. There were also questions regarding the potential for the rubber to release airborne particles and heavy metals. To address these concerns, a grant was made available to the New York Community Trust, and a study was performed to assess the potential release of airborne contaminants from artificial sports surfaces containing crumb rubber and the potential exposure of children to those airborne contaminants. The results of the study showed that there was no appreciable impact in the breathing zone from the 'contaminantes orgánicos persistentes' (persistent organic contaminants) in the crumb rubber and so a risk assessment was not warranted for the inhalation exposure route.

A project started by WRAP in 2000 [78] looked into the effectiveness of using 1–3 mm rubber crumb from waste tyres as a soil improver and a turf protector for a high-quality sports surface. The sports surface chosen was the rugby pitch at Cardiff Arms Park in Cardiff, Wales, which is home to Cardiff Rugby Football Club (Figure 7.6). The rubber crumb was used as the basis for the manufacture of a soil improver product with the trade name of Crown III, using a process patented by Tebbutt Associates. Crown III is a rubber crumb-based product used to improve the wear resistance of natural turf playing fields such as football, rugby and cricket pitches. The pitch, with an approximate playing area of 9,000 m^2, was treated in 2000 with around 60 tonnes of Crown III produced from 1–3 mm waste tyre crumb. This amount of crumb was the equivalent of around 8,600 waste car tyres. The pitch was inspected at regular intervals and, even after 6 years (the recommended time between treatments is four years), it was still benefiting from the original rubber crumb treatment with improved drainage and wear resistance. The use of waste rubber crumb as a soil improver for

Figure 7.6: Rubber crumb has been used as a soil improver and turf protector at Cardiff Arms Park rugby pitch in Cardiff, Wales. Reproduced with permission from the WRAP Quality Protocol, Waste and Resources Action Programme, Banbury, UK. ©WRAP (see note with Figure 7.1).

sports surfaces also provides additional environmental benefits because it has a reduced bulk density compared to the traditional soil improver – sand – so there is a reduction in net tonnage of treatment material per unit area of land improved. This reduces the amount of traffic required to bring the material on site. Other benefits include a reduction in player injuries due to the improved playing surface, improved growing rates and greater frost protection due to increased absorption of sunlight, and overall economic benefits, all of which offset the greater initial outlay (compared to using sand), in combination with the reduced maintenance demands and a harder wearing surface.

WRAP have also funded a project that looked into the use of cryogenically ground 0.5–1.5 mm rubber particles from waste car tyres [79] to help create a new artificial football pitch at Woodley Sports Football Club in the UK (Figure 7.7). The rubber particles were used in the FieldTurf 'third-generation' artificial sports pitch product, which is supplied worldwide by the American company FieldTurf Inc. 'Third-generation' artificial sports surfaces were developed in the late 1990s to address the problems encountered with earlier products. The ERDC Group is the sole supplier of FieldTurf surfaces in the UK, and was involved in this project. FieldTurf Inc. has a patent on the use of cryogenically ground rubber for their artificial sports pitches because the cuboid, relatively smooth particles help to keep the rubber in suspension with the sand and to shed water from the surface. In addition to the ground rubber, sand is also used to infill a plastic carpet with ~63 mm long pile to

Figure 7.7: The third-generation artificial football pitch at Woodley Sports Football Club, near Stockport, UK, contains cryogenically ground car tyre rubber. Reproduced with permission from the WRAP Quality Protocol, Waste and Resources Action Programme, Banbury, UK. ©WRAP (see note with Figure 7.1).

create the surface. Around 100 tonnes of rubber is required to produce a FieldTurf surface for a standard sized football pitch, which equates to around 150 tonnes of car tyres (21,000 tyres). Some of the advantages of using waste rubber in this way include the replacement of virgin silica sand as an infill material, and reduced maintenance costs, lower wear rates and fewer cancellations compared with natural grass. After two years of regular use, the pitch showed no signs of wear, even in the high activity areas such as the goal mouths. The feel of the surface and ball bounce characteristics were very similar to natural grass, and players did not suffer burns or other injuries when falling on the surface.

The subject of one of WRAP's 'Waste Tyres Case Studies' [80] was the use of waste rubber crumb from commercial vehicle (CV) tyres in PU binders for the surfacing of playground areas in the London Borough of Hammersmith and Fulham and the Magna Science Centre in Rotherham (Figure 7.8). These two WRAP-funded projects were carried out in 2005 by Playtop Ltd, a subsidiary of the Charles Lawrence Group plc, and the majority of the crumb used in each case was from CV

Figure 7.8: Playground areas in the UK have been surfaced using tyre crumb mixed with a PU binder. Reproduced with permission from the WRAP Quality Protocol, Waste and Resources Action Programme, Banbury, UK. ©WRAP (see note with Figure 7.1).

tyres, with the remainder being from sources such as post-industrial waste and, in some cases, virgin EPDM. The two playground surfaces were 346 and 550 m^2 in area and consumed approximately 15–23 tonnes of rubber, respectively. The surfaces were produced on site by blending the rubber crumb with PU and 'wet pouring' in two layers. The base layer is a coarse (20 mm mesh) rubber crumb from the recycled tyres and it is this that provides most of the safety features of the product. The top, wearing surfaces are typically 15 mm thick and mainly EPDM. Where a coloured top surface is required, colour virgin EPDM granules have to be used.

Playground surfaces of this type exhibit excellent characteristics in that they are stable, offer good protection against injuries caused by falls and trips and, in the event of damage, both base layer and top layer can be repaired relatively easily. The cost of the products depends very much on the top layer, with coloured play surfaces being around 60% more expensive than black ones. In the recycling process that Charles Lawrence Group plc uses for the waste CV tyres, the rubber granulate that is obtained represents 61% of the total weight of the tyres, with the remaining 39% (metal, fibre and so on) also being recycled, resulting in a 100% recyclable product.

A novel flooring product [81] that can be used in play areas for children and other areas where slip resistance properties are demanding has been developed that consists of a substrate of 100% waste tyre granules covered with a 6 mm coloured EPDM granulate top layer. The waste rubber and coloured EPDM granules are bound together using a flexible, one-part PU binder. Alternative base layers to the waste crumb with which the EPDM top layer can be combined include wooden floors, ceramic tiles or concrete.

A WRAP project [82] looked at using waste tyre rubber mixed with recycled aggregate and quarry 'fines' to construct a surface for a new bridleway (Figure 7.9). This usage route for waste rubber is of interest in the UK, where it is estimated that 3,158 km of rights of way are in need of some form of surface construction. The case study was conducted in Nottinghamshire in 2005 and 2006 and was the first time that such a surface had been produced in the UK. Three sections of the bridleway, which is sited at Clipstone, near Mansfield in Nottinghamshire, were identified and resurfaced using different construction material ratios. As these were used in conjunction with sections that had been resurfaced using standard bridleway construction materials, the opportunity was provided to obtain a direct comparison of their performance over an extended period of time. All three sections used rubber crumb derived from waste tyres at rates between 16 and 0.3 kg/m^2 of pathway, depending on the construction design criteria. In total, 8.5 tonnes of recycled tyre rubber were used, the equivalent of around 1,200 car tyres. The surfaces were monitored for durability and performance and, after four months of use, inspection indicated that the different constructions behaved differently.

Figure 7.9: Bridleways in Nottinghamshire, UK, have been resurfaced using compounds containing waste tyre crumb. Reproduced with permission from the WRAP Quality Protocol, Waste and Resources Action Programme, Banbury, UK. ©WRAP (see note with Figure 7.1).

7.6 Miscellaneous uses

Goettler and co-workers [83] have published the results of a research programme that looked at the reinforcement of blends of post-consumer NBR and HDPE with wood pulp and a silicate nanoclay. Compatibilising agents were used to improve the wetting and interfacial interaction of the reinforcement materials with the polymer matrix. Testing of the reinforced blends was carried out to obtain the physical, thermal and morphological properties, with the results showing that the wood pulp provided mechanical strengthening in tension, while the clay enhanced the stiffening and improved the tear strength.

Wu and Chen [84] blended reclaimed rubber powder from waste tyres with fly ash and a coupling agent (aminopropyl triethoxysilane) and investigated the physical and morphological properties of the resulting blends. In addition to investigating the influence that the amount of fly ash had on these properties, they also evaluated the effect of adding different cure systems (peroxide and sulfur types) and the temperature of cure. The results showed that the fly ash was an excellent filler and could be used as a replacement for silica fillers in reclaimed rubber powder composites of this type.

Wu and Zhang [85] looked into the possibilities of utilising waste rubber powder, pulp sediment and sawdust to produce composites. In these composites, when all three components were used, the pulp sediment formed the matrix, the waste rubber powder was the toughening agent, and sawdust was the reinforcing agent. The effect of varying the level of waste rubber on the physical properties of the composites was investigated, and their morphology examined using SEM and transmission electron microscopy. When all three components were used, the best recipe was found to be rubber 10 phr, pulp sediment 100 phr, and sawdust 30 phr. The results showed that when the rubber/pulp sediment ratio was 8:100, the samples had a smooth surface, high hardness, uniform arrangement and good compatibility. The mechanical properties of the rubber/pulp sediment blend were greater than those of the pure pulp sediment and of the rubber/sawdust/pulp sediment blend. The best curing conditions for the rubber/pulp sediment blends were found to be 15 min at 150 °C under a pressure of 5 MPa.

Yagneswaren and co-workers [86] have described how they used devulcanised rubber tyre powder as a filler in epoxy rubber composites. They investigated the effect that different loadings of the powder had on the chemical, thermal and physical properties of the resins and found that the tensile strength was enhanced as the loading of the powder increased to a level of 9 parts per hundred of resin, indicating that the powder was acting as a reinforcing filler. An investigation by the same team [87] looking at the effect of using devulcanised waste tyre powder as a filler in diglycidyl ether of bisphenol A epoxy resin showed that the presence of the tyre powder improved the heat resistance and curing characteristics of the composite compared to an unfilled epoxy of the same type. The local shear yielding within the matrix also showed that there was an effective interaction between the rubber and the epoxy resin.

In order to help create markets for waste rubber, WRAP has published a series of 'Waste Tyre Case Studies' that help raise awareness of the types of products in which waste rubber can be used. One of these focuses on the use of 2 mm waste tyre rubber crumb in the manufacture of carpet underlay (Figure 7.10) and was written in conjunction with the carpet underlay manufacturer Duralay Ltd, a subsidiary of the Interfloor Group [88]. In this case study, Duralay were reported in 2006 to use about 60,000 car tyres per week in their underlay product range, and the amount of underlay flooring produced in this way was about 4.55 million m^2 per year. This was equivalent to around 3.2 million car tyres annually, approximately 7% of the UK's waste car-tyres at that time. The processing route that is used consumes about 27,870 tonnes of tyres to create 17,000 tonnes of crumb, a conversion rate of about 61%, with the remainder being the steel wire cords and fabric. The advantages of using waste rubber in carpet underlay, in which it accounts for about 90% of the final weight of the product, are that it provides equivalent cushioning, noise reduction and heat insulation to other thicker underlay materials.

Figure 7.10: Duralay^{TM} carpet underlay products contain waste car tyre crumb. Reproduced with permission from the WRAP Quality Protocol, Waste and Resources Action Programme, Banbury, UK. ©WRAP (see note with Figure 7.1).

It is possible to use waste rubber as a garden mulch product to help suppress weeds in borders and flower beds [89]. A company called Dunweedin', based at Middlewich in Cheshire in the UK, has sold a mulch product from waste tyres for a number of years. The mulch is made using a two-stage process. Initially, a patented process is used to reduce the waste tyres into the mulch, then a second process involves the use of recycled water-based paints to colour them and package them up for sale in commercial outlets such as garden centres and DIY stores. Dunweedin' was able to increase production in 2009 when the PPP Group, based in Ballymena, Northern

Ireland, using a grant from Michelin, opened a new plant to take truck tyres from the nearby Michelin plant and process them into mulch to supply under licence to the company. This venture also enabled Michelin to make progress towards its own goals of energy conservation and the reduction of industrial waste to zero. Tyres that are not sent by Michelin to the PPP Group are sent to Charles Lawrence International, where the tyre waste is converted into surfaces for children's playgrounds and sports arenas. The advantages that rubber mulch has over more conventional mulch products (e.g., tree bark) are that it is inert, does not absorb water, should not need replacing, and it offers the option of a range of different colours, which bark cannot.

References

1. J. Karger-Kocsis, L. Meszaros and T. Barany, *Journal of Material Science*, 2013, **48**, 1, 1.
2. *A Review of the use of Crumb Rubber Modified Asphalt Worldwide*, Project TYR032-001, WRAP, Banbury, UK. *http://www.wrap.org.uk/*
3. D. Murphy, M. Wei, J. Mead, A. Cialone and M. Grubb in *Proceedings of the Annual Technical Conference*, Milwaukee, USA, Society of Plastics Engineers, Bethel, CT, USA, 2008, p.598.
4. Z-H. Zhu and C. Tzoganakis, *Journal of Applied Polymer Science*, 2010, **188**, 2, 1051.
5. S.M. Egodage, J.F. Harper and S. Walpalage, *Progress in Rubber, Plastics and Recycling Technology*, 2009, **25**, 4, 213.
6. Z.S. Ling, Z.Z. Xiu and K.J. Kuk, *Journal of Macromolecular Science: Part B*, 2011, **50**, 4, 762.
7. C-E. Wang, H. Chu and C-Q. Li, *China Rubber Industry*, 2009, **56**, 7, 393.
8. H. Ismail, D. Galpaya and Z. Ahmad, *Journal of Vinyl and Additive Technology*, 2010, **16**, 2, 141.
9. H. Ismail, D. Galpaya and Z. Ahmad, *Polymer Plastics Technology and Engineering*, 2009, **48**, 4–6, 440.
10. H.A.I.H. Dayang and R.A. Rahim in *Proceedings of the Thermoplastic Elastomers Conference*, Prague, Czech Republic, Smithers Rapra Technology Ltd, Shawbury, UK, 2008, Paper No. 11.
11. H. Ismail, D. Galpaya and Z. Ahmad, *Polymer Testing*, 2009, **28**, 4, 363.
12. *Scrap Tire News*, 2008, **22**, 3, 6.
13. M. Balasubramanian, M.A. Paglicawan, Z-X. Zhang, S.H. Lee, Z-X. Xin and J.K. Kim, *Journal of Thermoplastic Composite Materials*, 2008, **21**, 1, 51.
14. H.M. Da Costa, V.D. Ramos, W.S. Da Silva and A.S. Sirqueira, *Polymer Testing*, 2010, **29**, 5, 572.
15. L.S. Montagna and R.M.C. Santana, *Plastics, Rubber and Composites*, 2012, **41**, 6, 256.
16. M. Awang, H. Ismail and M.A. Hazizan, *Polymer Testing*, 2008, **27**, 1, 93.
17. C.D. McAfee and M. Grubb, *Rubber World*, 2008, **237**, 6, 31.
18. S. Anandlan, R.S. Rajeev, S.K. De, P.P. De and A.K. Bhowmick, *Kautschuk Gummi Kunststoffe*, 2009, **62**, 10, 529.
19. J. Qin, H. Ding, X. Wang, M. Xie and Z. Yu, *Polymer Plastics Technology and Engineering*, 2008, **47**, 1–3, 199.
20. H.M. Da Costa and V.D. Ramos, *Polymer Testing*, 2008, **27**, 1, 27.
21. Z. Hrdlicka, A. Kuta and J. Hajek, *Polimery*, 2010, **55**, 11–12, 832.
22. M.M. Abou Zeid, S.T. Rabie, A.A. Nada, A.M. Khalil and R.H. Hilal, *Polymer Plastics Technology and Engineering*, 2008, **47**, 4–6, 567.
23. Q-Y. Zheng, X-X. Zhang, C-H. Lu and M. Liang, *Polymer Materials Science and Engineering*, 2010, **26**, 7, 147.

24. O.P. Grigoryeva, A.M. Fainleb, V.F. Shumskii, V.A. Vilenskii, N.V. Kozak and N.V. Babkina, *Polymer Science Series*, 2009, **A51**, 2, 216.
25. S. Yagneswaran, N. Tomar, D.W. Smith and J.R. Cellura in *Proceedings of the ACS Rubber Divison Meeting – Fall 2009*, Pittsburgh, USA, American Chemical Society, Washington, DC, USA, 2009, Paper No.157.
26. Y-G. Li and L. Zhang, *Luntai Gongye*, 2010, **30**, 4, 221.
27. M. Guzman, N. Agullo and S. Borros in *Proceedings of the ACS Rubber Division Meeting – Fall 2009*, Pittsburgh, USA, American Chemical Society, Washington, DC, USA, 2009, Paper No.135.
28. H. Chandra and C.R. Pillai in *Proceedings of the ACS Rubber Division Meeting – Fall 2009*, Pittsburgh, USA, American Chemical Society, Washington, DC, USA, 2009, Paper No.63.
29. H. Chandra and C.R. Pillai in *Proceedings of the ACS Rubber Division Meeting – Fall 2010*, Milwaukee, USA, American Chemical Society, Washington, DC, USA, 2010, Paper No.67.
30. J.A. Shell in *Proceedings of the ACS Rubber Division Meeting – Fall 2009*, Pittsburgh, USA, American Chemical Society, Washington, DC, USA, 2009, Paper No.86.
31. T. Weber, M.G. Oliveira, M. Zenu, J.S. Crespo and R.C.R. Nunes, *Polymer Bulletin*, 2008, **61**, 2, 217.
32. N.F. Ushmarin, N.G. Vasil'yeva and N.I. Kol'tsov, *International Polymer Science and Technology*, 2009, **36**, 7, T/47.
33. M. Ahmadi and R. Yazdani, *Iranian Journal of Polymer Science and Technology*, 2008, **21**, 3, 217.
34. D.A. Baeta, J.A. Zattera, M.G. Oliveira and P.J. Oliveira, *Brazilian Journal of Chemical Engineering*, 2009, **26**, 1, 23.
35. N.N. Zulkepli, H. Ismail and A. Rashid, *Iranian Polymer Journal*, 2009, **18**, 2, 139.
36. N.Z. Noriman, H. Ismail, C.T. Ratnam and A.A. Rashid, *Polymer Plastics Technology and Engineering*, 2010, **49**, 1–3, 228.
37. N.Z. Noriman, H. Ismail and A.A. Rashid, *Polymer Testing*, 2010, **29**, 2, 200.
38. N.Z. Noriman, H. Ismail and A.A. Rashid, *Polymer Plastics Technology and Engineering*, 2008, **47**, 10–12, 1016.
39. N.Z. Noriman, H. Ismail and A.A. Rashid, *Polymer Plastics Technology and Engineering*, 2010, **49**, 7–9, 731.
40. D.H.A.I. Hassim and M.A. Ahmad, *Malaysian Rubber Technology Developments*, 2008, **8**, 2, 11.
41. M.V. Adov, P.L. Krasnov, S.Y. Pichkhidze and L.G. Panova, *International Polymer Science and Technology*, 2010, **37**, 6, T/47.
42. M.V. Adov, P.L. Krasnov, S.Y. Pitchkhidze and L.G. Panova, *Kauchuk i Rezina*, 2009, **6**, 32.
43. M.V. Adov, S.Y. Pichkhidze and K.L. Kandyrin, *International Polymer Science and Technology*, 2011, **38**, 12, T/25.
44. N. Roche, M.N. Ichchou, M. Salvia and A. Chettah, *Journal of Elastomers and Plastics*, 2011, **43**, 4, 317.
45. F. Pacheco-Torgal, Y. Ding and S. Jalali, *Construction and Building Materials*, 2012, **30**, 1, 714.
46. K.B. Najim and M.R. Hall, *Construction and Building Materials*, 2010, **24**, 11, 2043.
47. *Project to Conduct Pre-production Trials using Crumb Rubber in a Range of Concrete Blocks*, Project TYR0003-03, WRAP, Banbury, UK, 2006. *http://www.wrap.org.uk/*
48. *Full Scale Operational Trials Involving the Use of Recycled Glass in the Manufacture of Concrete Blocks*, Project GLA44-055, WRAP, Banbury, UK, 2006. *http://www.wrap.org.uk/*
49. A. El-Gammal, A.K. Abdel-Gawad, Y. El-Sherbini and A. Shalaby, *Journal of Emerging Trends in Engineering and Applied Sciences*, 2010, **1**, 1, 96.
50. A. Benazzouk, O. Douzane, K. Mezreb, B. Laidoudi and M. Queneudec, *Construction and Building Materials*, 2008, **22**, 4, 573.
51. D.L. Jevtic, D.M. Zakic and A.R. Savic, *Hemijska Industrija*, 2012, **66**, 4, 609.
52. S.B. Mortazavi, Y, Rasoulzadeh and A.A. Khavanin, *Iranian Polymer Journal*, 2010, **19**, 3, p.197.

53. R.Z.Shakhnazarli, N.Y. Ishchenko and A.M. Guiliev, *International Polymer Science and Technology*, 2009, **36**, 10, T/55.
54. G.L. Baumgartner in *Proceedings of the 4th Rubber Modified Asphalt Conference*, Akron, OH, USA, 7–8th May 2009, American Chemical Society Rubber Division, Washington, DC, USA, Paper No.2.
55. *Rubberised Asphalt Testing to UK Standards*, WRAP, Banbury, UK, 2008. *http://www.wrap.org.uk/*
56. S.H. El Ashry, L. Awad, M.A. El Zaher, E. Elkharashe and A.A. Bkheat, *Progress in Rubber, Plastics and Recycling Technology*, 2008, **24**, 4, 273.
57. A. Yousefi, A.A. Yousefi and M. Arjomand, *Iranian Journal of Polymer Science and Technology*, 2008, **21**, 3, 235.
58. *Tire Technology International*, 2012, July, 2.
59. *Waste Tyres Case Study – Rejuvoflex Micro-asphalt*, WRAP, Banbury, UK. *http://www.wrap.org.uk/*
60. *10,500 Scrap Tires in Rubberized Asphalt for California City Streets*, The Smithers Report, Ed., M.E. Fay, Smithers Scientific Services Inc., Akron, OH, USA. (Source: RubberWorld.com – Suisan City, CA, USA, 9th September 2008.)
61. *130,000 Tires being Recycled for Highway Projects Worth $11 Million*, The Smithers Report, Ed., M.E. Fay, Smithers Scientific Services, Inc., Akron, OH, USA, 2008, **21**, 3b. (Source: Ministry of Highways and Infrastructure, Saskatchewan, 3rd September 2008.)
62. A. Serna, M.D. Rio, J.G. Palomo and M. Gonzalez, *Construction and Building Materials*, 2012, **35**, 1, 633.
63. *Scrap Tire News*, 2008, **22**, 3, 7.
64. B. Dawson, *Rubber and Plastics News*, 2009, **38**, 24, 11.
65. *Replica Roof Slates*, Project TYR3-001, WRAP, Banbury, UK, 2007. *http://www.wrap.org.uk/*
66. E. Bormashenko, V. Goldshtein, R. Barayev, T. Stein, G. Whyman, R. Pogreb, Z. Barkay and D. Aurbach, *Polymers for Advanced Technologies*, 2009, **20**, 7, 650.
67. R. West, *Rubber World*, 2008, **238**, 3, 29.
68. J. Zhao, X.M. Wang, J.M. Chang, Y. Yao and Q. Cui, *Composites Science and Technology*, 2010, **70**, 14, 2033.
69. R. Maderuelo-Sanz, J.M. Barrigon Morillas, M. Martin-Castizo, V.G. Escobar and G.R. Gozalo, *Latin American Journal of Solids and Structures*, 2013, **10**, 3, 585.
70. A. Borlea, T. Rusu and O. Vasile, *Materiale Plastice*, 2012, **49**, 4, 275.
71. *Waste Tyres Case Study – Moulded Products: From Vibration Mats to Barricades*, WRAP, Banbury, UK. *http://www.wrap.org.uk/*
72. *Waste Tyres Case Study – Tile Adhesives: Flexibility from Tyres*, WRAP, Banbury, UK. *http://www.wrap.org.uk/*
73. *Potential for the use of Waste Tyre Derived Materials for Land Drains, Protection against Land Movement and use in Building Systems*, Project TYR0009-05, WRAP, Banbury, UK, 2006. *http://www.wrap.org.uk/*
74. *Warmer Bulletin*, 2011, **131**, April, 31
75. *Scrap Tyres – Basic Information*, US Environmental Protection Agency, Washington, DC, USA, 16 July 2011.
76. R. Uctas, *ICIS Chemical Business*, 2008, **274**, 5, 32.
77. *Scrap Tire News*, 2009, **23**, 5, 9.
78. *Waste Tyres Study – Cardiff Arms Park: Rubber Keeps Soil Green*, WRAP, Banbury, UK. *http://www.wrap.org.uk/*
79. *Waste Tyres Case Study – Artificial Turf Sports Pitches: Third Generation Artificial Pitch at Woodley Sports FC near Stockport*, WRAP, Banbury, UK. *http://www.wrap.org.uk/*

80. *Waste Tyres Case Study – Playtop Safer Surfacing: Tyre Crumb in Play Surfaces*, WRAP, Banbury, UK. *http://www.wrap.org.uk/*
81. *SA Plastics Composites and Rubber*, 2010, **7**, 6, 52.
82. *Waste Tyres Case Study – Bridle Ways: Tyre Rubber for a Better Surface*, WRAP, Banbury, UK. *http://www.wrap.org.uk/*
83. L.A. Goettler, M. Benes and M.H. Marko, *Composite Interfaces*, 2009, **16**, 7–9, 599.
84. W-L. Wu and D-J. Chen, *International Polymer Processing*, 2008, **23**, 2, 223.
85. W. Wu and J. Zhang, *Iranian Polymer Journal*, 2012, **21**, 11, 763.
86. S. Yagneswaran, N. Tomar, J.R. Cellura, G. Wallace and D.W. Smith, *Polymer Chemistry Division Preprints*, National ACS Meeting, Salt Lake City, USA, American Chemical Society, Washington, DC, USA, 2009, **50**, 1, 234.
87. S. Yagneswaran, N. Tomar, D.W. Smith, J.R. Cellura and G. Wallace in *Proceedings of the ACS Rubber Division Meeting – Fall 2008*, Louisville, USA, American Chemical Society, Washngton, DC, USA, 2008, Paper No.16.
88. *Waste Tyres Case Study – Duralay: Car Tyre Crumb used for Carpet Underlay*, WRAP, Banbury, UK. *http://www.wrap.org.uk/*
89. D. Fortson in *Wilde about Old Car Tyres*, Energy and Environment Section, *Sunday Times*, 5th April 2009.

8 Other methods for recycling and re-using waste rubber

8.1 Reclaiming process

As has been mentioned in Chapter 2, these are much older technologies, initially developed in the 19th century, that were only originally designed to 'reclaim' tyre rubber in some form for re-use as an additive in fresh compound. The three main processes of this type are as follows:

- *Thermal process*: Tyre pieces are steam-heated in a pan in the presence of air for around 2.5 h at up to 260 °C. The rubber product is then cooled, ground and consolidated on a two-roll rubber mill.
- *Heater or pan process*: A steam autoclave is used at 100–300 psi for an extended time with ground rubber mixed with reclaiming agents, such as 20% ammonium persulfate solution. After the process, the product is mixed with an aromatic oil and formed into sheets.
- *Digester process*: This is similar to the heater process, but a caustic soda solution is used. The ground rubber remains submersed in the solution and so less oxidative, chain-scission-type degradation occurs. The post-process regime includes washing, drying, grinding and consolidation on a two-roll mill.

An extensive review of the reclaiming processes used in the rubber industry up to 1962 was edited by Nourry [1]. A lot of work has been done to refine the original processes over the years and to understand the underlying chemical reactions that take place during them. Products that are manufactured using processes similar to those that are described above are still being sold commercially for use in the rubber industry as process aids and diluent additives, and they still have a role to play in the industry.

The more sophisticated methods of devulcanising rubber that are described in Chapter 4 can be looked upon as developments and improvements on the original 19th century reclaiming processes in that there are some basic common features, e.g., heat and chemicals.

8.2 Pyrolysis products

8.2.1 Introduction

At pyrolysis temperatures (e.g., 800–1,200 °C) waste rubber (usually tyres that are either whole or shredded) can be reduced down to gas, oil, char, metal and inorganic fractions that, with further processing, can then be used as energy sources, or additives and starting materials for other products.

https://doi.org/10.1515/9783110644142-008

An illustration of the products present in the gas, oil and char fractions resulting from the pyrolysis of a styrene-butadiene rubber (SBR) tyre is presented in the following [2]:

- *Pyro-gas fraction – 10–30% (contains ~1% sulfur)*: CO_2*, carbon monoxide*, methane, ethane, butadiene, propane, propene and butane, plus hydrocarbons (*dependent on the amount of oxygen that is available).
- *Oil fraction – 38–55% (<1% sulfur)*: high content of aromatic hydrocarbon compounds, as well as alkanes, alkenes, ketones and aldehydes.
- *Char fraction – 33–35% (~3–5% sulfur)*: carbon and ~15% inorganic ash (mostly zinc oxide).

Carrying out secondary processing steps on these three fractions enables them to be converted into value-added products. For example, once purified, the pyro-gas fraction can be used as an energy source to help run the pyrolysis operation. The oil fraction can be turned into carbon black by a furnace process, or into fuel oil, or chemical feedstock by distillation. The char fraction can be treated to yield products such as activated carbon, recovered carbon black and recovered inorganic compounds.

It is also possible to recover other important products from waste tyres when they are pyrolysed, one being the steel that goes into their construction. The approximate proportions of steel and other principal products obtained by this route are shown in Table 8.1.

Table 8.1: Approximate amount of each generic class of product from the pyrolysis of waste tyres.

Product	Approximate yield (%)
Fuel gas	10–12
Fuel oil	40–45
Carbon black	30–35
Steel wire	10–15
Inorganic material	>15

Some studies have looked at producing useful products, effectively fillers, from waste rubber by using methods that use a partial pyrolysis approach. Silica is being used as a part replacement for carbon black as a filler in the production of 'green tyres'. As this practice increases, there will be a corresponding increase in the total recoverable inorganic fraction from waste tyres, and so this recycling route could become more attractive. Examples of partial pyrolysis processes are described below.

Ivanov and Mihaylov [3] in Bulgaria have described how a solid product was obtained from 'green tyre' treads by a partial pyrolysis route, which, when analysed by infrared spectroscopy and X-ray fluorescence (XRF) spectroscopy, was found to have the composition shown in Table 8.2.

Table 8.2: Composition of the solid product formed by pyrolysis of green tyres [3].

Product	Amount (%)
Silicon dioxide	65
Carbon	30
Zinc oxide	3
Other components	2

This solid product was incorporated into SBR/butadiene rubber (BR) blends at various levels, and the mechanical and dynamic properties were compared with those of conventional silica/carbon black (2:1 ratio) filled SBR/BR compounds. The results showed that the physical properties, and the dynamic ones (such as heat build-up and tan δ), of the two types of compound were comparable. Their work showed that the recovered silica had a similar particle size to virgin silica, and its ability to disperse evenly within the rubber matrix was also similar.

Ivanov and co-workers [4] also pyrolysed scrap tyres in the presence of water vapour to produce a solid product. This product was then characterised by Fourier-Transform infrared and energy-dispersive XRF spectroscopy and evaluated for use as a filler in SBR cured compounds. These vulcanisates were then tested and the results obtained compared with SBR compounds that had been prepared using conventional fillers such as silica and mixtures of silica and carbon black

In another study [5], scrap rubber was pyrolysed to a limited degree and the product used as a filler in new compounds. In this case, tyres were pyrolysed by a low-temperature cracking process under slightly elevated pressures to generate a solid, powdery product. This powder was then incorporated at two levels (50 and 75 phr) into SBR and natural rubber (NR) compounds that were vulcanised and then tested to obtain properties such as heat ageing characteristics, mechanical strength and viscosity. In addition to the solid powder, this process also generated an oil product that had the potential to replace conventional hydrocarbon oils as a softener and extender in rubber compounds.

Some more detailed information on the fuel products obtained by pyrolysis is provided in Sections 8.2.2 and 8.2.3.

8.2.2 Pyro-gas and oil

With respect to the oil product, it can either be used unfractionated as heating oil, or can be fractionated into light, medium and heavy fractions. Possible applications for these fractions include:

- *Light fraction*: gasoline additive.
- *Medium fraction*: plasticiser for rubber.
- *Heavy fraction*: feedstock for coke production and a component of pavement surfaces.

When a completely non-oxidising atmosphere is used (i.e., vacuum pyrolysis), it is also possible to recover carbon black from the tyre rubber. This is called pyrolytic carbon black (PCB) and it has the potential to be used in new rubber products (Section 8.2.3).

A Vietnamese research group from the Ministry of Construction's Building Materials Institute has produced diesel oil from waste rubber [6]. The product was developed during a project on the thermolysis of rubber to help reduce environmental pollution caused by waste rubber products. The project used thermolysis technology to recycle waste rubber, including bicycle and automotive tyres, into industrial oil. Their investigations revealed that 1 kg of waste rubber produces around 0.4 kg of oil, and the group claims that the fumes produced from the furnaces do not pollute the environment, with the levels of nitrogen oxides, CO_2 and carbon monoxide meeting Vietnam's standards for atmospheric quality and industrial exhaust fumes. The research group transferred their novel technology to the Saigon Energy Industrial and Environmental Joint Stock Company, giving it a system capable of recycling 200 kg of waste rubber per hour.

8.2.3 Recovered carbon black

An important marketable product that can be obtained from the char fraction is recovered carbon black (also called PCB). The carbon black recovered from the pyrolysis of waste tyres accounts for around 30–35% of the total material generated by the process, with the exact figure depending on the types of tyres being pyrolysed.

Using the pyrolysis of waste rubber to obtain carbon black is claimed to be more environmentally friendly than obtaining it from oil using the usual manufacturing processes, such as the furnace or thermal route. As always, there are compromises, and the carbon black that is recovered does not have the same properties as the carbon black that was mixed into the rubber in the first place. There are a number of reasons for this, including the fact that a lot of rubber products contain a blend of two or more blacks to obtain the desired properties in the final product. The recovered carbon black will therefore be a mixture of these blacks and, as a consequence, its properties will reflect this with it having, for example, a wide particle size distribution. Even in cases where only one type of carbon black has been used, the recovered black normally has a different average particle size, particle size distribution, particle agglomeration (i.e., structure), surface chemistry and number of chemically active sites when compared to the original black. The large difference in surface chemistry results from the effect that the other constituents of the rubber compound (e.g., mineral fillers) have on the black as the matrix is pyrolysed away. Irrespective of these differences, the recovered black still has commercial value and is regarded as having an influence on rubber compounds similar to that of semi-reinforcing blacks [7].

An example of a company that is producing carbon black that has been recovered from waste tyres on a commercial scale is Carbon Clean Tech AG, which has had an industrial-scale plant operating since 2012. In a paper delivered to the *Tire Technology 2014* conference in Cologne, Yon [8] of Carbon Clean Tech AG explained the economic and ecological rationales of the process as being:

- Economic:
  - Raw material cost saving of 10–20%.
  - Independence of the oil price volatility.
- Ecological:
  - Tonne of oil saved per 1.0 tonne of rubber processed.
  - 1.4 tonnes of CO_2 emissions avoided per 1.0 tonne of processed rubber.
  - 99% recycling of scrap tyres.

In the same presentation, Yon explained that a large number of companies (>50) had investigated the possibility of reclaiming carbon black from tyres, but that very few had continuously produced any commercial product for the market place. Also, the fact that recovered carbon black (or alternative carbon black as Carbon Clean Tech AG refer to it) is not like conventional carbon black was underlined due to the fact that it is derived from a mixture of rubber tyres, containing a mixture of carbon blacks, the structures of which will have been altered in service, and that the tyres contain a large number of other chemicals that influence the surface properties of the recovered black.

In another paper presented at *Tire Technology 2014*, Terrade [9] addressed the question of whether pyrolysis black and virgin carbon black were competitive or complementary. In the paper, the processes for producing the two types of carbon black were summarised, as were a range of properties of the two materials (e.g., chemical composition, size, surface area, structure and surface activity of the particles). The overall conclusions were that pyrolysis carbon black can have a complementary effect on the activities of both the carbon black manufacturers and the tyre and rubber manufacturers. It was suggested that ways to help this would be for the three industries to integrate more closely and to form strong partnerships to promote the development and use of pyrolysis carbon black as a cost-effective 'green' alternative to combine with standard blacks in specially developed rubber products.

Cole [10–12] has carried out a thorough investigation into the performance of recovered carbon black as a filler in SBR compounds. A large number of processing, curing and physical property measurements were carried out, and the results obtained with one of these blacks (Phoenix Black E700) compared with those obtained on a comparable compound containing virgin carbon black.

A Chinese study [7] has involved subjecting PCB to demineralisation and then surface modification with a titanium coupling agent (NDZ-105), and then investigating how the black interacted with polyisoprene molecules by inverse gas chromatography. Rubber compounds were also prepared using these modified blacks and their mechanical properties were recorded. The results indicated that these blacks could be used to replace commercial semi-reinforcing carbon blacks in rubber compounds.

Du and co-workers [13] have carried out a study on PCB recovered from waste tyres. The scope of their work involved characterising the microstructure of the black using a scanning electron microscope (SEM) and then compounding the black into a number of ethylene- propylene-diene monomer (EPDM) compounds and producing cured samples. The influence of the PCB on the processing properties of the EDPM compounds was investigated, as was its effect on their mechanical and rheological properties. The results obtained were compared to EPDM compounds that contained the traditional fillers, furnace black N774 (a semi-reinforcing black), light calcium carbonate and thermal black N990. The SEM data revealed that the shape of the PCB particles was quite different to that of the N990 and calcium carbonate, but similar to the N774. Comparing the N774 and PCB with regard to other important particulate properties showed that the size of the primary particles of the PCB was smaller than the N774, but the particle aggregates of the PCB were larger. The processing properties of the EPDM compounds containing the four different fillers were found to be similar, but the PCB produced compounds that had a higher Mooney viscosity, and the surface appearance of the extrudate profiles was coarser. Compared to the other two blacks, the reinforcing effect of the PCB was found to be inferior to the N774, but similar to the N990. In common with other blacks, increasing the quantity of the PCB in the EPDM compounds was found to result in significant increases in tear strength and tensile strength.

Another group of Chinese researchers [14] also investigated and reported the effects on properties such as cure, viscosity and mechanical strength of incorporating PCB into SBR compounds, and compared the results with SBR compounds containing traditional carbon black fillers and calcium carbonate. A further paper by workers at Qingdao University [15] reported an elemental analysis of PCB, which showed it to comprise mainly carbon, oxygen, sulfur, zinc and silica, and that it had an ash content of up to 13.3%. This group compared the reinforcing properties of the PCB in SBR compounds with those of the traditional blacks N774 and N330, and found that the PCB was inferior to the N330, but similar in performance to the N774, and so could potentially be used as a partial or complete replacement for N774 in SBR compounds.

Cole and Spargo [16] compared the economic advantages of recovering and re-using the materials within tyres that have an intrinsic value (e.g., the carbon black) as opposed to using waste tyres as a source of tyre-derived fuel, the most popular disposal route in the USA, which only gives tyres a monetary value near to that of coal. They specifically reported on the attractive advantages of using the recovered black, Phoenix Black E900, in rubber compounds as a substitute for virgin black.

Other work that has looked into the performance of recovered black includes a Malaysian investigation [17] that evaluated the performance of the black in NR and nitrile rubber (NBR) formulations. The curing characteristics, physical properties and swelling index of the rubber compounds were investigated and compared with those of rubber compounds containing virgin black. In the case of the NBR compounds, blends of the recovered black with virgin black were also prepared and tested.

Scrap butyl rubber inner tubes represent a source of waste rubber that is easily identified and segregated. Workers in Pakistan [19] have subjected these items to a batch non-oxidative pyrolysis process under both uncatalysed and catalysed conditions. In both cases atmospheric pressure was used, with zinc oxide being used as the catalyst in the catalysed version. The resultant product in both cases consisted of a hydrocarbon mixture and carbon black. The effects of temperature, amount of zinc oxide and reaction time on the yield of the products were studied. The hydrocarbon fraction was assessed as a fuel by carrying out tests such as flash point and aniline point.

A Canadian company has developed a novel technology for recycling tyres into oil, steel and carbon black that uses microwaves in a nitrogen chamber [19]. Although the process is claimed to produce the same products as the pyrolysis route, it is said to achieve this at a lower temperature and so at a lower energy cost. In 2010 it was announced that a pilot plant was to be constructed in Ontario that would be capable of reclaiming 900 tyres a day, that another plant was planned for 2010/2011, and that there were plans to expand globally with a target of 6,000–7,000 tyres a day.

Montanari [20] has recently described how a Spain-based company, Curti Costruzioni Meccaniche SpA, has developed a batch pyrolysis plant process for waste tyres that is capable of producing a number of end products, e.g., syngas, oil and char. The char produced by this batch process has been analysed by XRF and the results obtained are shown in Table 8.3.

The char from the Curti process was also characterised by the Brunauer, Emmett and Teller (BET) method and the results compared to some conventional carbon blacks. The data obtained are shown in Table 8.4.

In addition to the batch process, Curti have finalised the design for a continuous pyrolysis plant to produce secondary raw materials and energy from waste

Table 8.3: Elemental data obtained on char product from the Curti process [20].

Product	Element (%)					
	C	H	N	O	S	Other
Char	85.5	3.4	0.4	4.6	2.7	3.4
New tyre[a]	86.9	7.6	0.4	2.9	1.6	0.6

[a] Literature values

Note: Heavy metals were below the detection limit of 5 mg/kg

Table 8.4: BET data obtained on char from the Curti process and conventional carbon blacks [20].

Product	BET value (m^2/g)
Char	63.7
N330	78.0
N351	70.0
N550	39.0

Note: The char from the Curti process was therefore found to have a similar BET value to N351 carbon black

tyres, which will be constructed according to a patented system. The capacity of the plant will enable it to process 10,000 tonnes of whole waste tyres per annum and will produce a number of products, for example:

- Syngas
- Fuel oil
- Carbon black and activated carbon
- Steel
- Energy (from the carbon products and oil)

The syngas that is generated by the pyrolysis plant will be used to power it. The plant is capable of dealing with whole tyres, so does not need any pre-grinding step, and can be customised to meet local requirements.

8.3 Depolymerisation to produce monomers and low-molecular-weight compounds

This type of recycling process to generate valuable by-products is not exclusive to waste rubber and can be applied to any hydrocarbon- based polymers. These processes have some similarities to the pyrolysis routes in that they result in the formation of low-molecular- weight molecules. The main difference is often that the pyrolysis process takes place in the absence of oxygen, whereas it can be present in these processes.

One approach described in the patent US 6,841,709 involves oxidative decomposition of the polymer by the use of chemicals such as nitrogen dioxide and/or dinitrogen tetraoxide in either an inert atmosphere or supercritical CO_2. This is a method that can be applied to any hydrocarbon-based polymer and in this way, a large range of polar mono- and difunctional chemical compounds can be produced that can be used as feedstocks for the production of other products, or as products in their own right. A second patent, US 5,516,952, partially oxidises polymers using a supercritical (or near supercritical) water mixture. With this method it is claimed that high yields of alkanes, alkenes, aromatics, alcohols, carboxylic acids and ketones, among others, can be produced.

Another type of depolymerisation process involves reducing the waste rubber from tyres back to its very basic chemical units of carbon monoxide and hydrogen. For example, a process for the conversion of waste tyre rubber into butadiene has been reported by GEM Fuels [21]. The process first converts the rubber into ethanol, which is oxidised to acetaldehyde, and then catalytically reacts with additional ethanol to generate butadiene. The butadiene can then be mixed with styrene in various proportions and copolymerised to produce virgin SBR, which has similar properties to the SBR used initially to produce the tyre.

A company called Coskata [22] has developed a process to make ethanol from practically any renewable source, including household waste, plant waste and waste tyres. The process involves three steps, the first of which thermally depolymerises the feedstock through gasification into its basic building blocks carbon monoxide and hydrogen (called syngas). Whereas syngas is normally then converted catalytically into alcohols, Coskata feeds it to anaerobic bacteria, which convert the carbon monoxide and hydrogen into ethanol. The third step in the process is then solvent recovery, where the ethanol is separated from the fermentation mixture. The process is claimed to have several advantages over traditional corn-based ethanol processes, including improved energy efficiency. It generates 7.7 times more energy that it consumes and uses about 1 gallon of water per gallon of ethanol. These figures compare well with those for corn-based ethanol production, which generates 1.3 times more energy and uses about 3–4 gallons of water per gallon of ethanol produced.

8.4 Generation of energy by incineration

The total energy present in waste tyres and other waste rubber products can be recovered by incineration. This process is initiated at high temperatures (e.g., 1,150 °C), is highly exothermic, and once under way is usually self-supporting. Used tyres have a calorific value of approximately 32,000 kJ/kg and the heat generated is used to produce steam that can then be used for purposes such as the heating of buildings or the generation of electricity. The electrical output has been estimated as being in the range of 20 megawatts for an installation with the capacity to incinerate 170,000 tonnes of tyres per year [23].

The degree to which energy is generated in this way varies in different regions of the world, often due to local considerations such as whether landfill regulations are in place and how much landfill space is available.

In the USA, a large proportion of waste tyres are incinerated, around 50% [24], but the practice is not carried out as widely in Europe. Some estimates of the proportion of tyres that are incinerated for a number of countries and geographical regions are provided in Table 8.5 [25].

Table 8.5: Percentage of waste tyres incinerated for energy generation [25].

Region/country	Percentage of tyres incinerated (%)
USA	53
Canada	20
Europe	41
Japan	70
Brazil	69
Mexico	0
South Korea	77
Australia	22
New Zealand	0

In order to address environmental concerns, however, efficient 'scrubbing' systems have to be employed to clean up the emissions (e.g., sulfur dioxide) from these types of plant so that their stack emissions meet the local environmental regulations that are in place.

Incineration is often carried out by rubber converters to produce their own energy and is popular for providing the energy to power cement kilns (see below). A comparative assessment of the amount of energy that is provided by tyres has been carried out, and the comparison with more traditional sources of energy is a favourable one, as shown in Table 8.6 [26].

Some environmental benefits may also result from the use of waste tyres for energy generation. A study funded by a number of large tyre companies and carried out by the French company Aliapur [27] has found that scrap tyres produce less CO_2 emissions when burned than solid fuels such as coke and coal.

Table 8.6: Energy contents of tyres, heavy oil and coal [26].

Product	Energy content (kcal/kg)
Tyres	8,000
Heavy oil	9,000
Coal	6,000

A large number of waste tyres are incinerated in cement kilns to produce Portland cement because they are particularly well suited to being a source of energy for this particular application. The reason for this is that the kilns operate at very high temperatures (e.g., 2,000 °C) and this converts the steel in the whole tyres to iron oxide as well as recovering the zinc oxide from the rubber and creating CO_2 and sulfur dioxide. The sulfur dioxide does not escape from the kiln, as it reacts with the calcium oxide present to form calcium sulfate. These salts and the metal oxides that are generated are useful ingredients in the final cement product [28].

References

1. *Reclaimed Rubber, Its Development, Applications and Future*, Ed., A. Nourry, Maclaren & Sons Ltd, London, UK, 1962.
2. M.A. Wojtowicz and M.A. Serio in *Conversion and Utilisation of Waste Materials*, Ed., M. Rashid Khan, Taylor and Francis, London, UK, 1996.
3. M. Ivanov and M. Mihaylov, *Journal of Elastomers and Plastics*, 2011, **43**, 4, 303.
4. M. Ivanov, M. Mihaylov and L. Ljutzkanov, *Kautschuk Gummi Kunststoffe*, 2010, **63**, 7–8, 303.
5. J. Magryta, C. Debek, R. Stepkowski and T. Kondlewski, *Elastomery*, 2009, **13**, 3, 21.
6. *Vietnam Research Group Produces Industrial Oil from Waste Rubber*, Ed., M.E. Fay, The Smithers Report, Smithers Scientific Services Inc, Akron, USA, 2008. (Source: VietNamNet – VietNamNet Bridge, 10th June 2008.)
7. J. Zhou, T. Yu, S. Wu, Z. Xie and Y. Yang, *Industrial and Engineering Chemistry Research*, 2010, **49**, 4, 1691.
8. M. Yon in *Proceedings of the Tire Technology Expo 2014*, 11–13th February, Cologne, Germany, 2014.
9. F. Terrade in *Proceedings of the Tire Technology Expo 2014*, 11–13th February, Cologne, Germany, 2014
10. W.F. Cole, *APRI Journal*, 2010, June, 13.
11. W.F. Cole, *Rubber World*, 2010, **241**, 5, 20.
12. W.F. Cole in *Proceedings of the ACS Rubber Division Meeting – Fall 2009*, Pittsburgh, USA, American Chemical Society, Washington, DC, USA, 2009, Paper No.97.
13. A. Du, Z. Zhang and M. Wu, Express, *Polymer Letters*, 2009, **3**, 5, 295.
14. D. Aihua, S. Changyan, C. Hong and L. Haiyan, *China Synthetic Rubber Industry*, 2008, **31**, 6, 446.
15. A. Du, M. Wu, C. Su and H. Chen, *Journal of Macromolecular Science B*, 2008, **B47**, 2, 268.

16. W.F. Cole and P. Spargo, *Rubber and Plastics News*, 2008, **38**, 2, 13.
17. D.H.A. Ismawl, *Malaysian Rubber Technology Developments*, 2010, **10**, 2, 31.
18. M.R. Jan, F. Jabeen, J. Shah and F. Mabood, *Journal of Thermal Analysis and Calorimetry*, 2010, **101**, 1, 303.
19. *Plastics and Rubber Asia*, 2010, **25**, 171, 1.
20. I. Montanari in *Proceedings of the Tire Technology Expo 2014*, 11–13th February, Cologne, Germany, 2014.
21. *Scrap Tire News*, 2008, **22**, 3, 10.
22. M. Bryner, *Chemical Week*, 2008, **170**, 10, 53.
23. M. Myhre, S. Saiwari, W. Diekes and J. Noordermeer, *Rubber Chemistry and Technology*, 2012, **85**, 3, 408.
24. *Scrap Tire Markets in the United States*, 9th Biennial Report, Rubber Manufacturers Association, Washington, DC, USA, May 2009.
25. *Managing End-of-Life Tires*, World Business Council for Sustainable Development, Geneva, Switzerland, November 2008.
26. N. Ishizawa, *International Polymer Science and Technology*, 1994, **21**, 1, T/60.
27. *Scrap Tire News*, 2009, **23**, 12, 10.
28. M. Myhre and D.A MacKillop, *Rubber Chemistry and Technology*, 2002, **75**, 429.

9 Conclusion

This book has described how environmental, legislative, economic and societal pressures have resulted in considerable worldwide efforts to develop effective and commercially viable methods for the recycling of waste rubber. These pressures, and the funding that has been made available by industry and government, have enabled a wide range of options for the devulcanisation of waste rubber to be developed and optimised, and a large range of viable possibilities are now available. A large number of these processes have enabled high-quality, added-value products to become available commercially. The studies that have been carried out in academia, and within industry, have demonstrated that these products can be used very successfully, in some cases without the need for blending with virgin rubber, to manufacture high-quality rubber products for the general rubber goods sector. In the tyre sector, devulcanised rubber is already being used in the manufacture of some new tyres, albeit at relatively low levels of addition. The research being carried out in this area continues to show promising results, offering the possibility that, in the future, higher levels will be used and that it will be used in a wider range of tyres than is the case at present.

In addition to devulcanisation, this book has shown that there are a large number of other possible routes available for the recycling and re-use of waste rubber. High-tonnage waste rubber products such as tyres continue to be used on a large scale to generate energy, and their conversion into fuel products, such as heating oils, and the recovery of carbon black for re-use, are well established and proven processes. This book has shown that one of the most important and economically viable routes for rubber recycling involves its conversion into powder or crumb and then introduction into a host of different matrices. Whether the host matrix is virgin rubber, or a thermoplastic to toughen it or manufacture a thermoplastic rubber, good results have been obtained. Good technical results have also been obtained when rubber crumb has been added to construction products like concrete, wood, ceramics and bitumen, or used in the manufacture of sports and recreational surfaces.

Although the technical argument can now be regarded as essentially won for recycling options that involve either devulcanisation or the use of rubber crumb, the principal factor that has often restricted commercial exploitation of both is the cost of the replacement product relative to the existing product. As has been demonstrated with rubber crumb, the use of it in, say, a road surface, often results in a product with better properties, but the cost has to be justified. With the devulcanisation of rubber for return into new rubber products, often the best that can be aimed for technically is equivalence with the existing virgin product, and the additional cost of the devulcanisation process can be a problem.

It is to be hoped – for both cases – that other factors such as increasing commodity and environmental costs and longer-term planning will assist in the

https://doi.org/10.1515/9783110644142-009

expansion of their commercial exploitation in the years ahead. These changes, together with continual improvements in the collection and sorting of waste rubber, and the setting up of closed-loop in-house facilities where tonnages allow, should ensure that the overall market for the recycling and re-use of waste rubber should continue to grow for many years to come.

Appendix 1 Brief overview of regulations applying to waste rubber in the European Union and the USA

A1.1 Introduction

A number of references to the regulations that apply to waste rubber and waste rubber products, such as tyres, have already been made during this book. As with all areas where regulators are taking a keen interest, the situation is complex and changing quite rapidly, so only a very brief summary will be provided here. It is, however, an area that is driving a lot of research and interest in the recycling of waste rubber and one that is of great importance in achieving successful commercial routes for its re-use. It was therefore thought important that at least some mention should be made of the more salient points that come up in typical enquiries involving the subject.

Accordingly, some specific examples of the regulatory situation in Europe and the USA are covered here. Some references to the developments that have taken place in other parts of the world such as China and Russia have already been made in Chapter 3.

Fortunately, for those that require information and a high level of detail in this area there is usually a comprehensive list of the applicable up-to-date regulations, and available guidance documents, on the national, European Union (EU) and US Food and Drug Administration websites.

A1.2 Situation in the United Kingdom

A1.2.1 Is end-of-life tyre rubber regarded as a waste product or not?

In the UK, tyre-derived rubber materials will normally be regarded as having ceased to be waste, and therefore no longer subject to waste management controls, provided they require no further processing before use, namely:

- They have been produced using only those input materials specified in Section 2 of the UK Waste Resources Action Programme (WRAP) Quality Protocol [1] (i.e., source segregated waste tyres).
- They meet the requirements of an approved standard or specification (e.g., Publically Available Specification (PAS) 107:2007).
- They meet the requirements of engineering standards if specified for the specific end-use.
- They meet any additional requirements specified by the customer.

https://doi.org/10.1515/9783110644142-010

- They have been produced using either ambient or cryogenic processing technologies.
- They have been processed to one of the size categories and in accordance with one of the options in Section 2 of the Protocol [1]; and are destined for use in one of the designated applications within the designated market sectors listed in Section 4 of the Protocol [1].

Producers (i.e., the operators undertaking the shredding, crumbing and granulating) have to demonstrate that the above criteria have been met and they can do this by obtaining certification from an approved certification body (e.g., BSI in the UK) and maintaining records.

A1.2.2 Waste shipment regulation (EC/1013/2006)

Tyre-derived rubber materials meeting the UK WRAP Quality Protocol in the UK may cease to be waste in England and Wales, but the country of destination may take a different view and, under the Waste Shipment Regulation, if the competent authority in that country considers the material to be waste then the controls specified in EU Regulation EC/1013/2006 will apply to the shipment.

A1.3 Situation in the European Union

A1.3.1 European waste catalogue

The European Waste Catalogue (2002) provides a comprehensive list of waste codes and descriptions based on waste source and type.

A1.3.2 REACH treatment of waste materials

In general, waste is exempted from the Registration, Evaluation, Authorisation and Restriction of Chemicals (REACH) Regulation (EC 1907/2006) as it is covered by separate waste management controls. However, once waste has been fully recovered and ceases to be waste, waste management controls cease to apply and REACH may apply instead at that point. Unless specifically exempted (e.g., because a substance has already been registered), producers may need to register substances recovered from waste and placed back on the market and make available appropriate hazard and safety information, for example a suitable safety data sheet.

A1.3.3 REACH treatment of aromatic process oils in tyres

The EU REACH regulation restricts the marketing of certain tyres or parts of tyres containing aromatic oils beyond a certain limit during their production. The limitation on aromaticity of oil in vulcanised rubber compounds is specified at 0.35% bay protons as measured and calculated using International Organization for Standardization (ISO) 21,461. This can also be referred to as polyaromatic hydrocarbon (PAH) content. Most tyre companies in the USA are meeting these specification requirements for the European market [2]. Manufacturers in other parts of the world are also altering their formulations to comply with this regulation. This means that it would be permitted to put waste rubber from these products back into new tyres. Problems may occur where the rubber originates from waste tyres that are more than around 10 years old, or the age or source is unknown, and so could contain relatively high levels of high-PAH-content oils. Fortunately, this potential problem will reduce in likelihood as time goes by.

A1.3.4 REACH treatment of devulcanised rubber, polymers, preparations and blends

If the waste rubber that has been devulcanised is known to have been manufactured in the EU, or to have been imported into the EU in compliance with REACH, and any chemicals that are used to devulcanise it are registered, than its composition should be compliant with REACH:

- Devulcanised rubber can then be sold in the form of masterbatches for blending into new rubber compounds or, occasionally, other polymers such as thermoplastics.
- Preparations (e.g., masterbatches) do not need to be registered. However, each substance that goes into the preparation needs to be registered by the manufacturer or importer if the amount manufactured or imported exceeds 1 tonne/year.
- Polymers are exempt from REACH, so if they are used as binders in masterbatches, as the base materials in polymeric products or in blends, they do not need to be registered.

A1.4 Situation in the USA

The Identification of Non-Hazardous Secondary Materials That Are Solid Waste rule published by the US Environmental Protection Agency (EPA) in 2012 determines whether non-hazardous secondary materials are considered fuels or wastes when burnt [3]. Combustion units that burn non-hazardous secondary materials that are classified as fuels are regulated as boilers under Section 112 of the Clean Air Act, while units that burn materials that are classified as wastes are regulated as incinerators under Section 129 of the Clean Air Act. In the rule, the EPA identified two secondary materials (resinated wood and scrap tyres managed under the oversight of the established tyre collection programmes) as non-wastes when used as fuel, and also identified off-specification tyres managed under the contractual relationship as non-waste fuels. However, scrap tyres that are discarded in stockpiles, landfills or monofills will be considered wastes unless these tyres are processed as specified in the rule.

In July 2011 the EPA finalised the Deferral for CO_2 Emissions from Bioenergy and Other Biogenic Sources under the Prevention of Significant Deterioration and Title V Programs rule. Under this rule, greenhouse gas (GHG) emissions associated with the NR fraction in tyre-derived fuel were exempt from EPA GHG permits for three years. At the end of this period the EPA will consider the GHG emissions that resulted from biogenic sources during it in order to properly account for them in the future.

The Toxic Substance Control Act (TSCA) requires that companies that are producing carbon black and other chemicals as a result of the pyrolysis of tyres have to report them to the EPA. The specific requirements of this reporting are laid down in the TSCA.

References

1. *WRAP Quality Protocol for Tyre-derived Rubber Material*, November, 2009, WRAP, UK. http://www.wrap.org.uk/.
2. PTA News Bureau, *Reclaim to Gain Higher Usage*, Polymers and Tyre Asia, 2011, April/May,80.
3. M. Myhre, S. Saiwari, W. Diekes and J. Noordermeer, *Rubber Chemistry and Technology*, 2012, **85**, 3, 408.

Abbreviations

ACS	American Chemical Society
ASTM	American Society for Testing and Materials
BET	Brunauer, Emmett and Teller
BR	Butadiene rubber
BS EN	British Standard European Norm
BTRC	Big Tyre Recycling Corporation
CBS	Cyclohexyl benzothiazole
CEN	Comité Européen de Normalisation
CR	Chloroprene rubber
CV	Commercial vehicle
DCPD	Dicyclopentadiene
DIN	Deutsches Institut für Normung e.V.
DIY	Do-it-yourself
DMA	Dynamic mechanical analysis
DPDS	Diphenyl disulfide
DPG	Diphenyl guanidine
DR	Devulcanised rubber
DRC	Devulcanised rubber compound
DSC	Differential scanning calorimetry
EC	European Community
ENB	Ethylidene norbornene
ENR	Epoxidised natural rubber
EPA	US Environmental Protection Agency
EPDM	Ethylene-propylene-diene monomer
EU	European Union
EV	Efficient vulcanisation
EVA	Ethylene-vinyl acetate
FKM	Fluorocarbon rubber
FRP	Fine reclaim powder
FTIR	Fourier-transform infrared
GHG	Greenhouse gas
GPC	Gel permeation chromatography
GRG	General rubber goods
GTR	Ground tyre rubber
HD	Hexadiene
HDA	Hexadecylamine
HDPE	High-density polyethylene
HNBR	Hydrogenated nitrile rubber
HSM	High-stress mixing
ILP	Island Leisure Products
IR	Infrared
ISO	International Organization for Standardization
LCA	Life cycle assessment
LDPE	Low-density polyethylene
LLDPE	Linear low-density polyethylene
MB	Masterbatch(es)

https://doi.org/10.1515/9783110644142-011

MBT	Mercaptobenzothiazole
MBTS	Mercaptobenzothiazole disulfide
MFI	Melt-flow index
Mn	Number average molecular weight
Mw	Molecular weight
NBR	Nitrile rubber
NMR	Nuclear magnetic resonance
NR	Natural rubber
ODR	Oscillating disc rheometer
OTR	Off-the-road
PAH	Polyaromatic hydrocarbon
PAS	Publically Available Specification
PCB	Pyrolytic carbon black
PE	Polyethylene
PET	Polyethylene terephthalate
PP	Polypropylene
PRL	Polymer Recyclers Ltd
PU	Polyurethane
R&D	Research and development
RA	Rubberised asphalt
RBR	Butyl rubber reclaim
REACH	Registration, Evaluation, Authorisation and Restriction of Chemicals
rNBR	Recycled nitrile rubber
RNR	Natural rubber reclaim
RSM	Response surface methodology
SBR	Styrene-butadiene rubber
SBS	Styrene-butadiene-styrene
SCRC	Self-compacting rubberised concrete
SEM	Scanning electron microscope/microscopy
SSSE	Solid-state shear extrusion
TARRC	Tun Abdul Razak Research Centre
TDF	Tyre-derived fuel
TGA	Thermogravimetric analysis
TMPTA	Trimethylolpropane triacrylate
TMTD	Tetramethyl thiuram disulfide
TPE	Thermoplastic elastomer(s)
TPV	Thermoplastic vulcanisate(s)
TS	Technical specification
TSCA	Toxic Substance Control Act
UHP	Ultra-high pressure
UV	Ultraviolet
VB	Vacuum bottom
VOC	Volatile organic compound(s)
WAG	Welsh Assembly Government
WRAP	Waste Resources Action Programme
XPS	X-ray photoelectron spectroscopy
XRF	X-ray fluorescence
ZMBT	Zinc mercaptobenzothiazole

Index

https://doi.org/10.1515/9783110644142-012

www.ingramcontent.com/pod-product-compliance
Lightning Source LLC
Chambersburg PA
CBHW061210240926
51846CB00006B/1170
9783110644005